MW01285537

After having tried two patterns, the Canadian Army has chosen this "computer made material" in 1999. (Lt. Bennett-DGPA-ISNT Ottawa, Canada)

Camouflage Uniforms
of European and NATO Armies 1945 to the Present

J.F. Borsarello

Schiffer Military History
Atglen, PA

Book Design by Ian Robertson.

Library of Congress Catalog Number: 99-65695

Printed in China.
ISBN: 0-7643-1018-6

We are interested in hearing from authors with book ideas on related topics.

Published by Schiffer Publishing Ltd.
4880 Lower Valley Road
Atglen, PA 19310
Phone: (610) 593-1777
FAX: (610) 593-2002
E-mail: Schifferbk@aol.com.
Visit our web site at: www.schifferbooks.com
Please write for a free catalog.
This book may be purchased from the publisher.
Please include $3.95 postage.
Try your bookstore first.

In Europe, Schiffer books are distributed by:
Bushwood Books
6 Marksbury Road
Kew Gardens
Surrey TW9 4JF
England
Phone: 44 (0)208 392-8585
FAX: 44 (0)208 392-9876
E-mail: Bushwd@aol.com.

Try your bookstore first.

Contents

Foreword

It can be read in some books of military studies that camouflage uniforms are mainly an expression of nostalgia about highly colored armies of the old times and napoleonic garments. A French general, having read my trials in this field, told me that elite troops wanted absolutely to be distinguished and proud of their condition, far from Signal corps, Supplies and even ordinary Infantry (nevertheless named "Queen of battles"). When I asked this general why so many people were wearing a camouflage suit, he replied that enemy shot first elite troops and officers. So everybody had to be camouflaged, even cooks and nurses! When I asked him why so many countries wore such garments the "anti camouflage" general replied that during maneuvers, different countries would fight together and their officers had to recognize their soldiers among the foreign troops! When in the end I told him again that animals in the bush, in savannas and in the depth of the seas could not know of napoleonic heavy colored uniforms and wore nevertheless camouflage covers, the general replied that I was an annoying person and, more than that, an impolite officer without respect for an old general covered in glory! By the way, certain armies wear camouflage suits out of war, indeed, mainly in parade and military demonstrations. A certain nostalgia of old glorious armies exists also, but there are many other reasons to improve the protection of men in war, and history proves it. How would one otherwise explain the millions of dollars spent to research the best colors and the best material in the U.S. and other countries, and the urgent request of MacArthur in 1942 for his "Marines" in the Pacific war, and the German general Hausser for his troops invading the USSR?

Camouflage uniforms are not a fashion, a nostalgia, a joke, or an unusual fact. Bright minds and artistic observers of nature have researched an insufficient power of human eyes, and a means to avoid detection. With optic illusions, a defect of colored vision, it is possible for a soldier to disappear in forest, desert, or rocks. To achieve one's ends, the researcher tests many melting colors, blurred drawings, faded tonalities, and different lights and shades. Many materials are rejected before the good one for a certain background or landscape is found. This requires much money, and many countries' armies cannot pay research centers. In this case the countries imitate—more or less—a good pattern discovered by a richer army, or buy the pattern. Nowhere in the world does there exist at present time a research center comparable with the U.S. Natick center. During the thirties such a center existed in Germany, from where issued kilometers of camouflage trials for Wehrmacht and Waffen SS, the "kings" of camouflage uniforms in the forties! So it can be seen that after the war, even former enemies have imitated German patterns. After the victory of the U.S. and Allied armies, many countries have imitated the famous pattern of camouflage uniforms of Pacific war "Marines."

After 1945, at the time of the Korean war (1950-1953) and Indochina conflict, the idea of camouflage was brought back. Belgium, France (and also the U.S. army in Korea) tested printed materials. The Warwaw pact and NATO tested camouflage uniforms from that time on, and in 1955, more than twenty patterns were worn by European armies, during the famous "Cold War." Most of these printed materials would be dropped and replaced by other, more increasingly efficient patterns, until 1970. Then in Vietnam, the U.S. army manufactured a pattern, called M65 and later M67 ERDL, which would be imitated by many armies in Europe and in Asia. But we had been living through the last years of effectiveness by this type of protection. Until now, the eye was the only way to discover the enemy. Tomorrow, with special materials, such as the M16A1 with TVS, the Hughes thermic camera in the U.S. army, night day visors Trijicon ALOG or the French Sopelem SFIM ODS night visor, or alternatively other instruments, the human eye will no longer suffice. The enemy will be discovered because of his thermic radiation up to 100 meters! His camouflaged uniform will become a museum piece and an object for collectors.

Only small armies, lacking financial means and without these sophisticated weapons, will again wear the old classic camouflage suits in jungles and woodlands. These armies are in large numbers, and will apply this excellent mean of protection. The camouflage clothes will be one day abandoned by military units and will take a large place in hunting, fashion, and even fishing!

Acknowledgments

Andrys Jérôme: specialist of textiles, Paris, France.
Bjelos Henad: Yugoslavian specialist of Uniformology.
Bousquet Didier: French collector.
Bouzigues Jean Pierre: Scientific Attaché in Embassy of France in Sarajevo, Bosnia Herzegovina.
Clausen Henrik: Danish collector, Copenhagen.
Champagne Roger: Headquarters of Canadian Army, Ottawa.
Czermelji (colonel): Czech Army in Prague.
Debay Yves: International photo reporter, Paris, France.
Delmoitez Palmer Celeste: NATO photo service, Brussels, Belgium.
De Scheel Lennart: Attaché of Defense in Embassy of Denmark, Paris, France.
Duschene A. (colonel): Headquarters in Luxembourg.
Edmond Blanc Cyrille: French collector of uniforms and insignias.
Garcia Servert Ruben: Attaché of Defense in Embassy of Spain, Paris, France.
Gillet Jean-Pierre: French collector.
Heidecker (colonel): Attaché of Defense in Embassy of Austria, Paris, France.
Hladik Miroslav (major): Ministerstvo Obrany Ceske Republiky, Prague.
Horgan Justin: Irish collector.
d'Ianni (captain): Headquarters in Rome, Italy.
Jérome Regis: French collector.
Kajshoj Anja: Attaché of Defense in Embassy of Denmark, Paris, France.
Karlaukis Aldis: Latvijas Kara Muzejs, Riga, Lettony, and Embassy in Paris.
Maçak Kamil (colonel): Ministry of Defense in Bratislava, Slovakia.
Mac Namara Brendan: Embassy of Ireland, Paris, France.
Mahlamaki (commander): Finnish Navy, Helsinki.
Marzetti Paolo: Italian collector.
Mérienne Patrick: specialist of flags, France.
Mjell Gunnar: Attaché of Defense in Embassy of Norway, Paris, France.
Milovanovic Zoran: photographer in Yugoslavia.
Mollo Andrew: British specialist of European uniforms.
Moskric Jose: specialist of uniforms, Lubliana, Slovenia.
Murray Ron: ISO Publications, London, United Kingdom.
Oksanen Markku (captain): Ministry of Defense, Helsinki, Finland.
Peterson Daniel: American specialist of camouflage uniforms.
Peucelle J.: French collector.
Prevezer Michael: British collector, London, United Kingdom.
Purins Artis: Ministry of Defense, Riga, Lettony.
Rellin Loren: specialist of uniforms, Culver City, USA.
Renoult Bruno: specialist of uniforms, Paris, France.
Rudberg Sven (captain): Swedish Navy, Stockholm.
Smith Digby: specialist of armies uniforms, United Kingdom.
Sostoï G.(major): Headquarters, Budapest, Hungary.
Soulier Jean-Pierre: collector, Lausanne, Switzerland.
Surgailus G.: Ministry of Defense, Vilnius, Lithuania.
Toni (colonel): Headquarters in Rome, Italy.
Tudman Ankica: Zagreb, Croatie.

I thank my wife, Marie-Pierre, for her large participation in this book, and my son, Nicolas, who helped me greatly with this work.

1

History of Camouflage in European and U.S. Armies

World War I

From 1914 to 1918, the idea of a camouflage uniform for soldiers seemed to be a matter of minor importance in comparison with the serious events of this period. Nevertheless, all armies in Europe at this time wore a khaki or gray plain material, except the French army, with its famous red "garance" trousers.

Some people thought that these colors are the symbol of military and represent a kind of tradition. Others preferred a less visible suit for fighting. Many trials were made, and famous painters proposed camouflaged clothes, such as André Mare, Dunoyer de

The famous jacket hand painted in 1915 by the French soldier Guingot. Probably the first printed material for camouflaged uniforms at this time. P. Mignot photo. Courtesy of Musée Lorrain of Nancy, France.

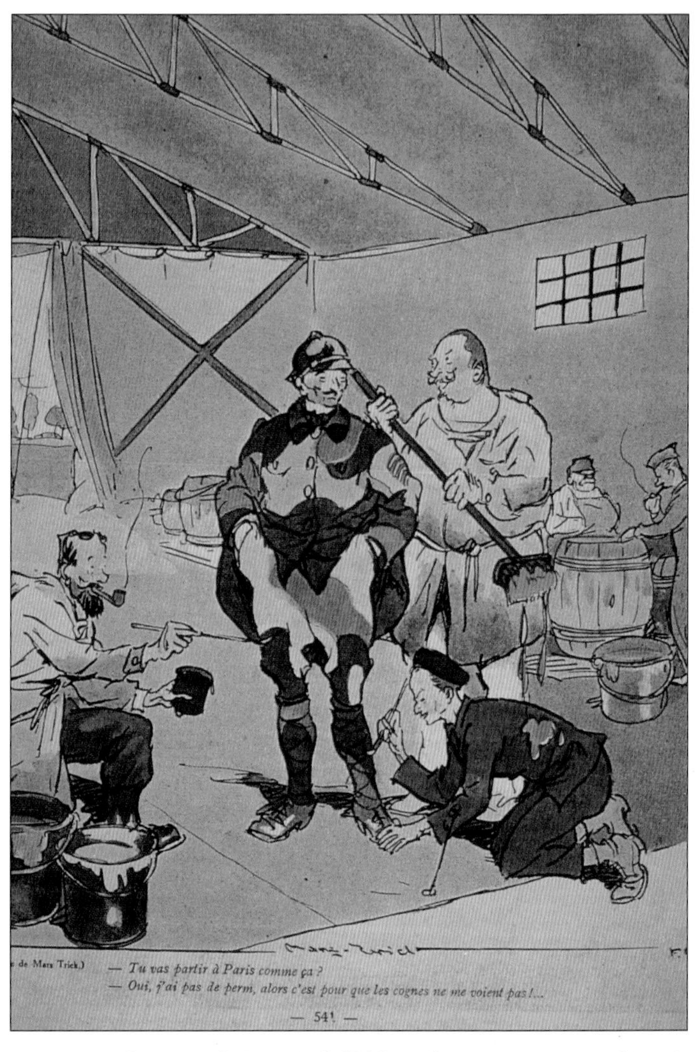

Humoristic drawing of Mars Trick 1915: "Why are you painting his uniform?" "He is going to Paris and does not have the permission of the captain. This way military police cannot see him." Courtesy of Musée of Péronne, France, 1915. La Baïomette Magazine.

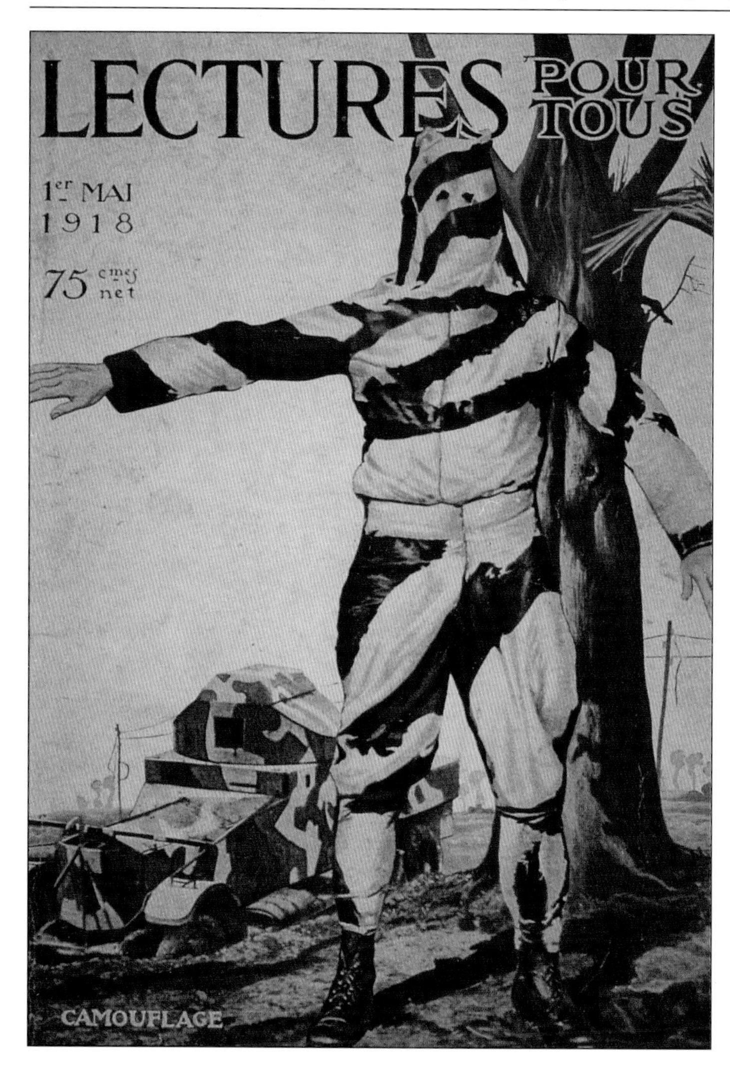

Front cover of the magazine Lectures pour Tous, 1st May 1918. Fanciful drawing of camouflage, soldier and armored car. Courtesy of Musée of Péronne, France.

Italian Ethiopian war in the thirties. A camouflage collective tent made of four quarter shelters. The 1929 pattern was probably the first especially printed in a factory for war equipment. Courtesy of Albertelli Magazine Uniformi, Parma Italy.

Segonzac, and Guirand de Scevola, but their drawings would stay in forgotten files. Only vehicles, armored cars, ships, and planes would receive camouflage colors. The famous "section of camouflage" would be employed by thousands of soldiers and civilians to fabric snares, such as false trees, false houses, and even false cannons in order to deceive enemies, as well as to camouflage tarpaulins to hide vehicles.

The red trousers (1914-1915) were probably the origin of many casualties, and they would be dropped and replaced by "bleu horizon" (horizon clear blue). This pattern, not very efficient in comparison with German "feldgrau" would remain, nevertheless, the symbol of the victorious French army in 1918. The khaki color would be adopted later, many years after the end of the war. The French army had been a pioneer, but a pioneer that had been stopped too early.

Between 1918 and 1940

The second army to have manufactured a camouflage pattern was Italy. The printed material was first used on tarpaulins, but soon after on ground shelters or tents in the famous 1929 pattern. Nobody knows why the drawings were so repetitive on such a short length and did not offer any camouflage efficacy. In nature, elements are always different and do not repeat every twenty inches! But it was the first time in which an army created such a printed material in order to protect men.

World War II

In 1939 the German army studied many early patterns, so various that in 1940, twelve sorts of drawings were tested in Poland, and later in France. At the end of the war in 1945, German WH and WSS wore more than twenty patterns, so efficient that soon after

Waffen SS Viking division wearing typical "palm trees and clumps" camouflage jacket in 1943. Private Collection.

Paratroopers of German Luftwaffe in "Afrika" 1943. Inspection of General Ramcke. ECPA Archives, France.

USSR camouflage uniforms 1939 worn by female soldiers of the Signal Corps. Novosty Agency, Paris, France 1978.

First trials by the U.S. Army of camouflage uniforms at Fort Ethan Allen, Vermont, in 1942. Courtesy of Signal Corps photo, U.S. Army.

The King Kard pattern HBT 1942 retained by HQ as a camouflage uniform and ordered for emergency use in the Pacific by General MacArthur.

many were imitated by foreign armies. English forces had been obliged to wear khaki since the wars in India and South Africa at the beginning of 20th century. In 1939, men were only provided with rubberized ponchos, with long brown and green waves, exactly like the camouflage on British planes. They were later issued the famous "Denison smock" worn by paratroopers units, which also remained as a symbol of the British army in WWII.

Few nations in war from 1940 to 1945 wore a national camouflage uniform: the USRR used two, one with long "fingers" brown on spinach green background, and another, green background also but sprinkled with various sorts of leaves, white or clear yellow (1943). Hungary used two sorts of quarter shelters, one brown with curious pink bowls and the other an olive color background with white or brown shapes.

British army uniforms for WWII, camouflage heavy tank crew pattern, Denison smock and light material for jungle. J. de Fromont photo, Musée of Pourrain, France 1995.

Belgian soldiers of the Korean battalion of General Crahay in 1951. Courtesy of Royal Military Museum, Brussels, Belgium.

General Salan in Indochina war with officers dressed in U.S. King Kard HBT given by the U.S. Army in 1947.

Post War and "Cold War," NATO

Later, many countries have chosen the U.S. M65-69 pattern because this sort of garment had been tested with success during the Vietnam war. Years before, French armies had success with the brush strokes lizard, and the Vietnamese imitated their pattern with the "tiger stripe," which in turn was imitated by the U.S. army in Vietnam for special forces. More than three hundred patterns existed between 1970 and 1985 in the world's armies, and many countries have a particular pattern for each special arm, while others prefer to have only one uniform for all units.

After the Fall of the Warsaw Pact

During Gulf war, Iraq used no less than twenty different patterns. In Africa, one can often see a group of soldiers each man has a different pattern....it is a real patchwork!

The camouflage idea, formerly reserved to military actions, has invaded civilian circles. Police wear "urban" camouflage, made of gray, black, and white shapes to blend with the concrete in urban areas. Sometimes, during night operations for surveying drug traffic, men of narcotic bureaus wear subdued blue garments to melt into night areas. Even frog men, nevertheless out of woodlands, will soon be painted like fish, and two or three navy units have camouflage special suits for amphibious fighters.

Even fashion has used drawings of camouflage; women can be seen dressed in famous patterns like Barracuda, King Kard, ERDL, though mixed with different colors. Is it possible to imagine such nice women disappearing among greens and walls? It would be a paradoxical action, and every man in the world would be desolated....

2

Camouflage Uniforms of Western Europe

Austria

In 1999, Austria is the only one country to not have any camouflage uniform. In the sixties, a reversible half shelter had been created, and later a two piece suit, a winter parka, bag, cover helmet and many accessories. The pattern was made of little spots of pink, brown, and green, but the general color, seen from a distance, appeared to be pink. These clothes were not reversible, but the half shelter could be used on two sides: one generally pink like the uni-

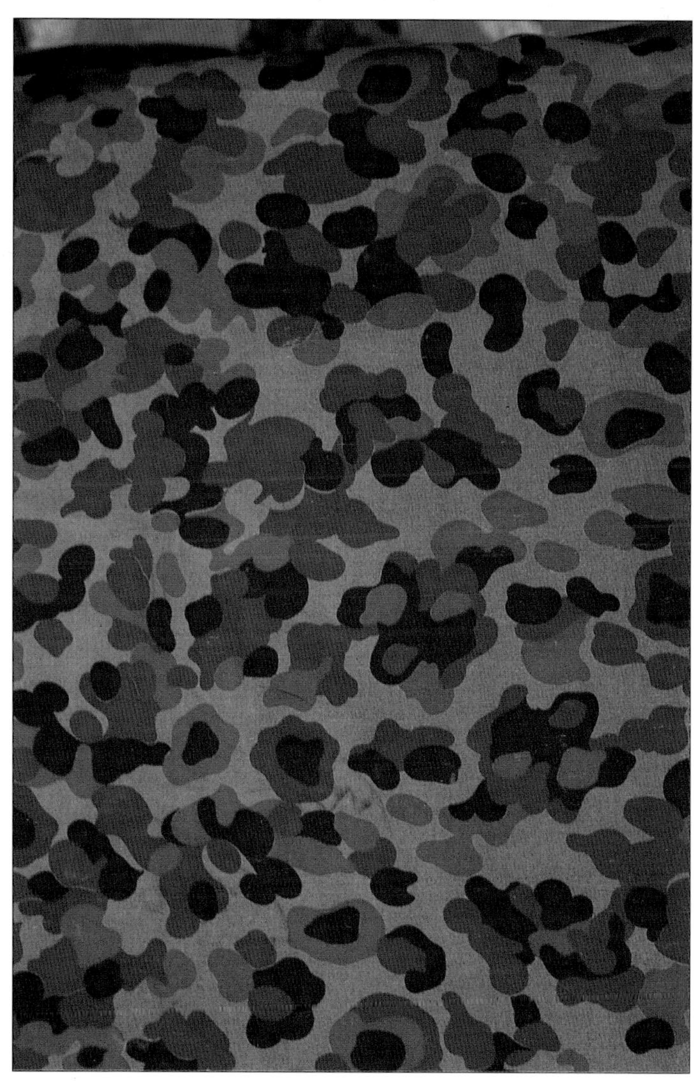

Close up view of woodland "pink" side of the camouflage pattern quarter shelter, which is reversible. The uniform itself is pink pattern and not reversible. Author's collection.

LEFT: *Reservist of the Austrian army wearing obsolete uniform 1962-78 in 1992. Weisshaupt Verlag photo, Graz, Austria.*

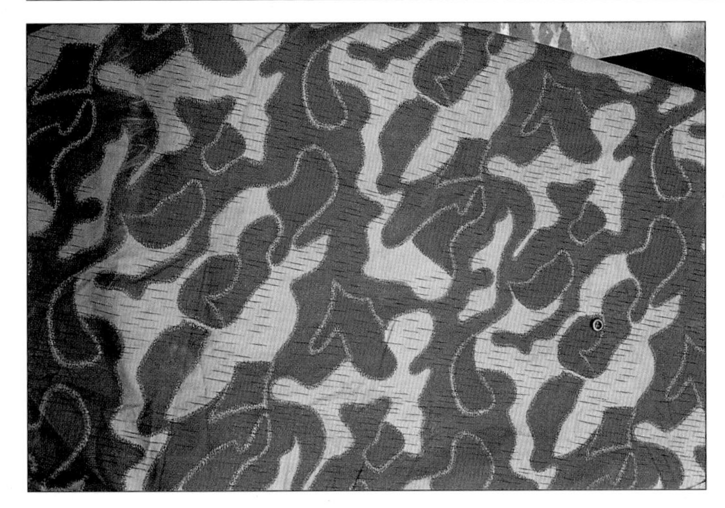

Close up view of autumn-rock side of the quarter shelter. Author's collection.

Camouflage jacket with hood and heavy padded winter material. Author's collection.

forms, and the other side with a pattern made of gray shapes, khaki green spots, and rain strokes. It can be said that the pink side was similar to German camouflage M44, called a "pea pattern," worn at the end of the war by Waffen SS troops. The rain strokes were specifically from the German camouflage 1943 pattern. In 1978, the Austrian army dropped all the camouflage suits, used only by reservist soldiers in maneuvers. Very curiously, "frog men" or commandos wear a diving rubberized one piece camouflaged suit. The pattern used for these special units is the same as the "amoeba" German army trial of the sixties.

Curious Austrian diving suit for "frog men" commandos. The pattern is the German 1960 "amoeba" printed on rubberized material for diving suits. Weisshaupt Verlag photo, Graz, Austria.

Belgium

The Belgian army has certainly been the country most concerned with camouflage trials in Western Europe. Indeed, eight trials have been tested between 1954 and the present.

The first uniform was worn by the famous Korean battalion in 1951 and two patterns have been used: moons and balls pattern, round shapes on khaki background, and a waves pattern, with long brown waves. Both patterns were made of heavy and rough material, probably for the hard climate of South Korea in winter. It can be said that very few armies of UNO wore camouflage uniforms in Korea in 1951.

In 1954-55, Belgium tested a very special pattern, created by the German army in February 1945. It was called the "leibermuster," and was designed for combat in rubble, ruins, and built up areas, and at the end of the war in destroyed cities. The Beligan pattern is exactly the same as the German pattern, but printed on rough and heavy material. The reverse of the material is, curiously, ochre red.

Later, a new material appeared, sort of a "moons and balls," but longer shapes with brush strokes very frayed. Called "large waves and brush strokes," this pattern was probably made by Utexbel in Renaix (Belgium) and sold later in Africa and Middle East. Two sorts of colors were used: green or khaki background.

At present time, almost all units of the Belgian army wear the "jigsaw" pattern, also called the "jigsaw puzzle" because the pattern is made of brown and green shapes fitted together. The background and shapes have variant colors in rain proof plastic quarter

Bugles of Belgian soldiers in 1951 wearing one of the first camouflage suits made of dark green and brown round shapes as "moons and balls." Courtesy of Royal Military Museum, Brussels, Belgium.

Colonel Crahay of Korean Battalion of Belgium with another pattern at trial, long brown waves. Courtesy of Royal Military Museum, Brussels

Close up view of one of the first Belgian pattern "moons and balls," 1952. Author's collection.

Close up view of 1954 Belgian trial "leibermuster."

Belgian pattern similar to German 1945 "leibermuster" made for urban combat in ruins and rubbles. Markings ABL 1955 KH LP 1710. 30. 11. 54. Soon dropped. Retained by Switzerland in 1962. Author's collection.

Belgian "large waves and brush strokes, dark green and brown shapes." Worn during humanitarian operations in African wars by the soldiers of general Laurent, paratrooper battalions. Private collection.

Close up view of Belgian large waves and brush strokes pattern 1960-75. Author's collection.

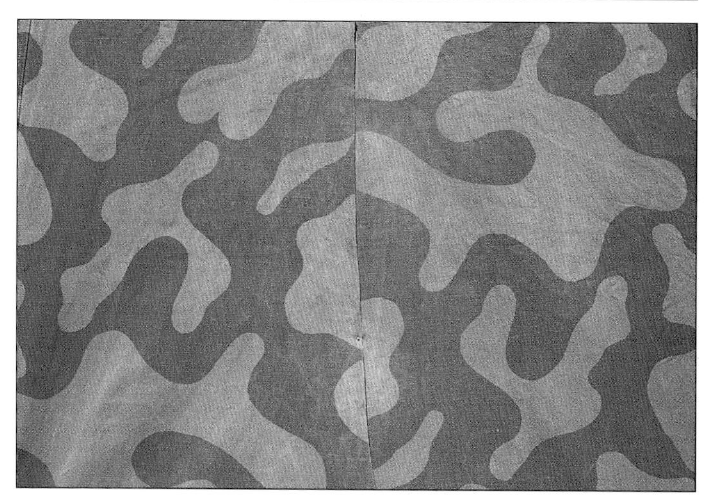

Belgian army long brown waves on gray background heavy material of 1950, dropped in 1960. Author's collection.

Female corporal driver of a military truck with up to date camouflage jig saw material. ECPA Archives, France.

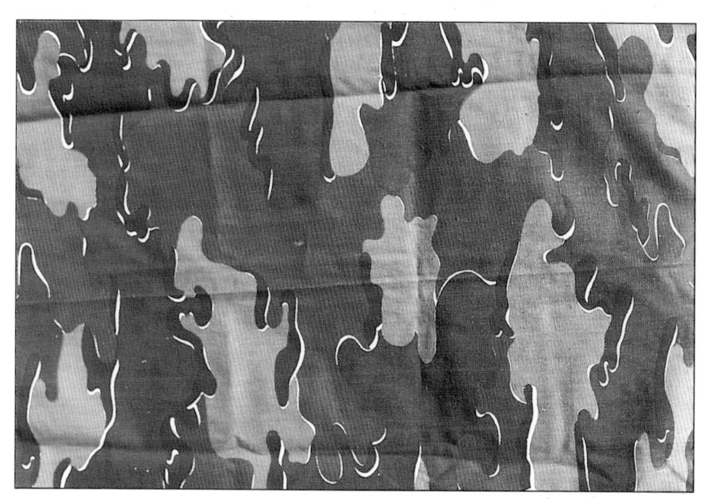

Close up view of jig saw pattern of Belgian army in 1998, mustard color background. Author's collection.

Waterproof material worn as jacket and often worn in maneuvers by Dutch soldiers or as a poncho in rainy periods. Private collection.

At left first 1951 camouflage "lizard" used in Indochina: clear colors, clear background. At right "FUMACO" fusiliers-marins-commandos, probably British origin (frog skin). Worn in Indochina and Algeria until 1963.

Paratrooper wearing the 1952-54 lizard pattern. Brush strokes are dark green and dark brown, olive drab background. Dragoon Anvil association, Nice, France.

shelter, or rain proof jackets often worn also in the Dutch army. Note that this material has been lightly modified for Zaire, Burundi, and Rwanda troops.

The recent "jigsaw puzzle" was seen in April '94 during the "silver back" operation, worn by two commando "Codos" paratroopers. It was also seen in March '93 during Somalia UNO action, in company "ESR" in Slavonia (a province at the north of Croatia) at the end of '95 with the 12th Prince Leopold Regiment, and in the 4th regiment of "chasseurs à cheval" (without their horses).

France

After 1945, the French army did not wear any national camouflage pattern until 1951. In the Indochina war, camouflaged suits were imported from the U.S. army and British forces in the Pacific. These uniforms were the King Kard US 44 pattern and the GB SAS, or colonial, light camouflage material. Only one unit would be clothed with a special pattern, never seen elsewhere, the "frog skin" worn by FUMACO (fusiliers marins commandos), who would wear this pattern again in the Algerian war until 1962. Nobody knows at present time the exact origin of this "frog skin," though maybe it was England.

In 1950, the French uniform commission decided to introduce a special camouflage for TAP (airborne troops). These first trials, made of "lizard" brush strokes, would be tested during maneuvers in 1952, under command of general Ridgeway, in Germany. Many variants would be made between 1952 and 1956, the camouflage uniforms being used in Algeria, and after some days during the Suez canal operation.

French soldier in Algeria 1958-62 with 1953 "Bigeard cap" and Mat 49 machine gun. Dragoon Anvil association, Nice, France.

After the end of the Algerian war, the French camouflage suit was quickly dropped, because the lizard pattern became synonymous with "colonial war." The decision to withdraw it was somewhat influenced by the involvement of some troops in a coup. By 1964, only French foreign legion soldiers wore "lizard" camouflage during exercises in the Sahara. But until 1999, many African armies wore this famous pattern, sold very cheap because of large stocks of supply units.

French soldiers in Algeria with "lizard" pattern and camouflaged quarter shelters. ECPA Archives, France.

Until 1990, no camouflaged uniforms could be seen in the French army. Certain generals were convinced that such material was unnecessary, "just an old nostalgia of napoleonic wars." But during these years, a curious "jigsaw puzzle" pattern was commonly worn in the French army—no regulation, but tolerated for maneuvers—in air forces commandos, 1st infantry regiment, and other units. The material was rainproof plastic with superimposed green and brown shapes. In 1981 Texunion France factories proposed a new camouflage pattern after meetings with German specialists of textiles, Marquardt and Schulz. The little spots have a very good camouflage efficacy, but the French army would refuse it. The reasons were the reluctance of headquarters and the great similarity with future German camouflage pattern 1980.

When the Gulf war started, the French army was the only army to join Saudi at the end of 1989 in a green olive drab uniform. But just before the first shot, camouflage for desert areas arrived just in

French soldiers in Suez Canal Operation in 1956 with an Egyptian soldier. They wear the 1954 "lizard" pattern. ECPA Archives, France.

French army in operation of Tchad. Note collective tents made of quarter shelters in desert areas pattern 1952, clear sand background and brush strokes. Private Collection.

Camouflage lizard in Algeria "commandos de chasse" 1959. ECPA Archives, France.

time, a very good pattern, very efficient. Later in 1991 a Central European pattern was proposed, and adopted for all units. At last it seems that the French army now accepts the principle of camouflage. Every soldier now wears the famous pattern: cooks, bureaucrats, truck drivers, and even dogs of units of survey.

So the pattern is seen everywhere: 11th airborne division, 9th RCP, 1st RHP, units of Almadin operation in Central Africa, naval commando in Rwanda, infantry battalion 6 in Bosnia 1996, 3rd and 8th RPIMA (overseas elite troops), and operational assistance troops (EFAO). Rarely worn now is the US M81 blouse style, as in the 13th dragoons paratroopers, for example. CRAPS commandos wear a

RIGHT: 1981 new trial for a national pattern, which was perhaps too similar to the 1978 German trial Marquardt and Schulz pattern. Author's collection.

1963-1989 period. Non-regulation saw tooth pattern, often worn in this time, waterproof plastic, made by K Way and also by Salik International, dropped in 1990. At right, first trial of Texunion factory 190. Not adopted.

French soldier during a "gun show" in Satory display, with the non-regulation pattern. Author's collection.

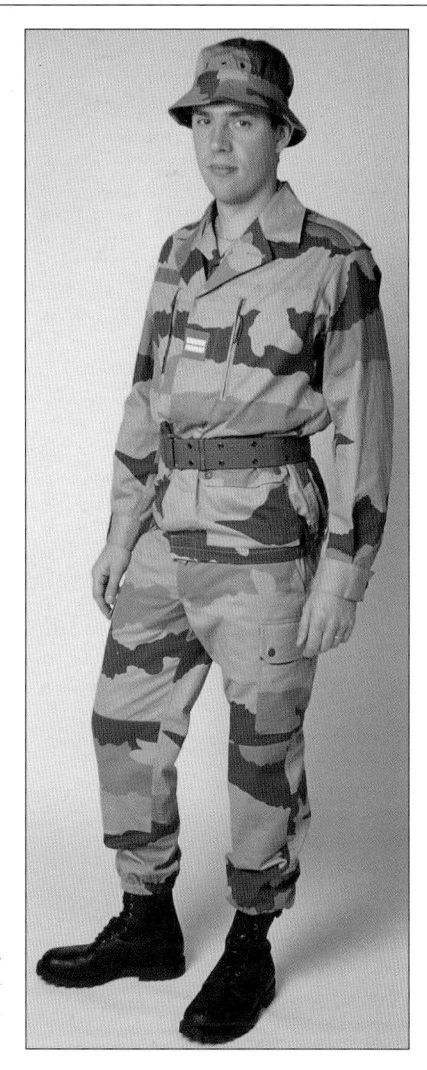

New desert areas pattern for French army. Distributed just in time for the beginning of the Gulf war.

Close up view of desert areas 1989. Two piece suit with cap or hat.

Central Europe pattern. The desert areas pattern has been worn by the Daguet division, 1st Combat Helicopter Regiment (1RHC), 1st RIMA ("Marines") in Sarajevo, 2nd RIMA elsewhere in Bosnia, 35th artillery regiment, and the 1st regiment of Spahis (old name of Moroccan and Algerian cavalry).

Everywhere in France, soldiers in camouflage uniform survey transports of civilians, even in the Metro (underground) in Paris. Many people afraid of terrorism are thus reassured, and certain generals say that the camouflaged uniforms, very visible in the crowd, are an element of peace. Moreover, the soldier so clothed is himself…revalued.

Officers and soldiers wearing the new desert areas pattern near the end of 1989 in the Gulf war. SIRPA Terre Archives, France.

"Central Europe" camouflage suit for French army in 1994. Same drawings as desert areas pattern but dark green instead of clear sand color, and moreover black twigs. Horizontal shapes for the two patterns.

Close up view of Central Europe pattern. Author's collection.

Germany (Federal Republic of Germany)

Interestingly, German WWII patterns continued to influence camouflage uniforms in Europe after the war, with variations of WH and WSS patterns used even in certain armies of North Africa, and eastern countries of the Warsaw pact, such as Poland, Bulgaria, and Czechoslovakia.

In West Germany in 1950, curious patterns were created, very similar to WH 1940-45, such as splinters and fluffy edges splinters, including a new sort of brown and khaki pattern with rain strokes, but were soon dropped.

In 1960 a really new pattern appeared, never issued in Germany, the "amoeba pattern," which was only used as a half tent. This pattern was reversible, one side for spring-summer and the other for autumn-winter, in the good tradition of German system of reversible material 40-45.

Between 1961 and 1974 two sorts of patterns were tested that were very different from former styles, the "dot pattern," made for close up or far distance camouflage, and the "ragged edged leaves," which was tested for a short time and never retained.

In 1976, at last, appeared the definitive pattern of the West German army, created by Marquardt and Schulz company, which would be very efficient and distributed until 1979 to all units. For mountainous landscape, a special snowy weather pattern was been created at the same time, made of white background and dark green pine tree clumps.

In 1989, German army added to its patterns collection a sand desert camouflage, not yet tested in a desert war.

East Germany, in the Warsaw pact, initially had a Russian camouflage suit never used in USSR and perhaps made only for the DDR army: ochre brown shapes like the "amoeba," and sorts of drawings similar to the first Russian camouflage 1937-40.

Later, in 1960, there appeared a gray pattern sprinkled with large ragged edged leaves, turquoise blue, light green and brown, but this pattern was not very efficient and would be soon dropped

1950 West Germany "splinters" similar to Wehrmacht material 39-45 (but not exactly the same). Author's collection.

1943 WH similar pattern for German Bundesgrenschutz (border guards). Author's collection.

New German army 1950 pattern in "maneuvers," West Germany. Private collection.

West Germany reversible Spring-Autumn pattern for quarter shelters, half tents, and cover helmets. Here summer side. Note splinters in background 1960. Author's collection.

Autumn side of 1960 pattern, for half tents (this pattern was used for the Austrian "frog men" diving suit).

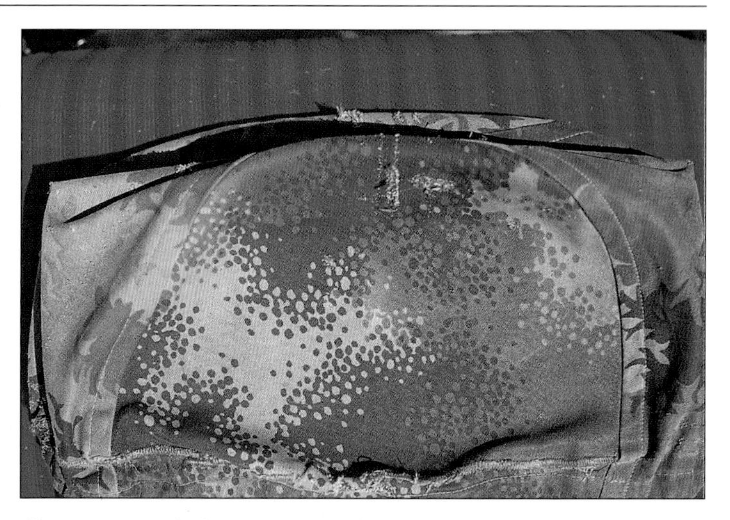

Close up view of "dot pattern." Author's collection.

A variant of splinters, but brown drawings, thick rain strokes and khaki background. Author's collection.

1974-76: "points" pattern, rarely seen made for "far away camouflage": dot pattern. Author's collection.

1976 "Saw Tooth" trial for West Germany with ragged edges leaves. Soon dropped. Courtesy of Dan Peterson, U.S. army in Germany.

RIGHT: 1976-99 German Marquardt and Schulz pattern now for all units. Author's collection.

German soldiers in 1998. NATO maneuvers. Courtesy of Ministry of German defense and Military Attaché of embassy in Paris.

Snow camouflage with pine tree needles in clumps for German mountain troops. U.S. cavalry magazine photo, USA.

German camouflage for desert areas 1998. Courtesy of Dan Peterson, U.S. Army in Germany.

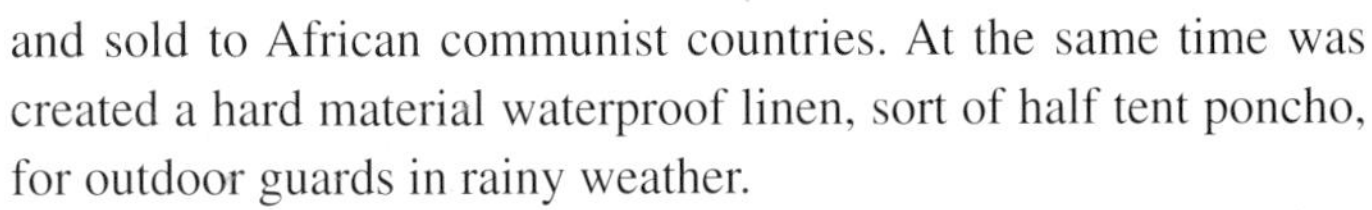

and sold to African communist countries. At the same time was created a hard material waterproof linen, sort of half tent poncho, for outdoor guards in rainy weather.

The last pattern, which was made in order to be worn by all armies of the future Warsaw pact, was used from 1975 until the fall of the Berlin Wall. It was made of a gray background sprinkled with classic rain strokes.

The recent Marquardt and Schulz, modified, could be seen in the reunified Germany in the Franco German brigade in Bosnia Herzegovina, the 110th regiment infantry in Gorozdne, and the 552nd jäger battalion, for example. Made of special plastic, the Marquardt and Schulz pattern has been created for NBC protection.

Ireland

The Irish army has sometimes worn, though rarely, the British DPM 58-78 pattern for certain maneuvers or police operations in the north of the country. During more than twenty years, the olive drab uniform had been used, this until 1997. In 1999, a new Woodland camouflage is worn similar to the French 1996 or U.S. Woodland 1981, but clear green shapes are a little "jig saw," or "saw tooth" made.

Italy

The Italian army was the first in the world to have printed, in 1929, the first camouflaged quarter shelter, with non-figurative drawings, known as the "telo mimetico." This pattern would be used until 1996 (almost seventy years)! Drawings were repeated too—every 50 centimeters—and presented bad camouflage efficacy, but it was the first in the world. Used mainly as a half shelter, in 1937 it became one of the first camouflage jackets for paratroopers. Throughout the years, colors have been modified, but the drawings were always the same.

In 1945 there appeared a new pattern, blue background, ochre brown long waves, only reserved to "Marines," specifically the famous battalions of San Marco. Later, only in 1968, the Italian army modified the pattern and chose a sort of leaf pattern very similar to USM65-67 EDRL of the U.S. army. The San Marco battalion would

East Germany 1949 pattern made for Neue Volks Armee (NVA). Author's collection.

Saw tooth pattern for East Germany made of turquoise, green and brown leaves on a gray background. Heavy material. Author's collection.

soon drop their old pattern to change to a pink camouflage suit made of fluffy and soft edges, and drawings which are somewhat similar to the German Sumpfmuster 1943. For the Gulf War, the Italian army had anticipated a desert areas pattern, issued in 1988: sand color background and brown or green twigs, which are the same as the woodland type but with a different background.

These uniforms have been seen in "Forza di intervenzione rapide," brigade Friuli, 185th regiment, 9th assault battalion, 9th infantry regiment in Bari, 15th "Stormo," and in Arena exchange in 1993. The desert area pattern has been worn during the Farfadet operation, in Somalia with San Marco battalions, and brigade Folgore in Sicily. The special San Marco pattern could also be seen during Vespri Siciliani in January 1993.

Luxembourg

The Luxembourg army has worn the U.S. M81 Woodland pattern since 1968-1975 for all units.

A NVA soldier with saw tooth leaves pattern (flachentarnmuster).

East Germany 1955 heavy material camouflage rain proof quarter shelter. Author's collection.

The last issued NVA camouflage uniforms "rain strokes" on gray background for all units and all materials until the fall of the Warsaw pact. Author's collection.

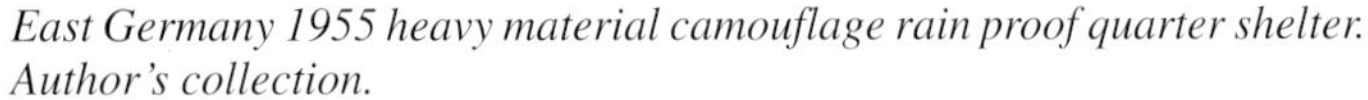

Netherlands

The Dutch army wore in 1946-49 in Indonesia a pattern very similar to U.S. M44 King Kard HBT. Currently, the Dutch army wears both British DPM and USM81 Woodland. They have been seen in Stoett Troppen, 1st battalion of "Mariniers," in March 1994 in Bolatice maneuvers in Bohemia, company 108 of commandos, and the 11th mob. Brigade. Recently, maybe at trial, there was seen the Belgian "jig saw puzzle" pattern, but with different colors, yellow or clear brown background, and also, modified drawings, a jig saw more stretched.

Portugal

In 1965, Portugal created a camouflage pattern for troops in Angola, where the war for independence had started. The first pattern was clear background with large dark brown brush strokes, and pale clear green ones, used at this time by "brigada aerotransportada

RIGHT: NVA tankists in "rain strokes" uniform in maneuvers under the Warsaw pact. Na straje i mira i socializma. Planeta.

The first camouflage of the Irish army was olive drab pattern with long linen strips, without any drawings. Courtesy of Mr. Mac Namara, Irish embassy in Paris.

Variant of picture for new camouflage pattern of Irish army.

The new camouflage of the Irish army will soon be distributed. Courtesy of Mr. Mac Namara, Irish embassy in Paris.

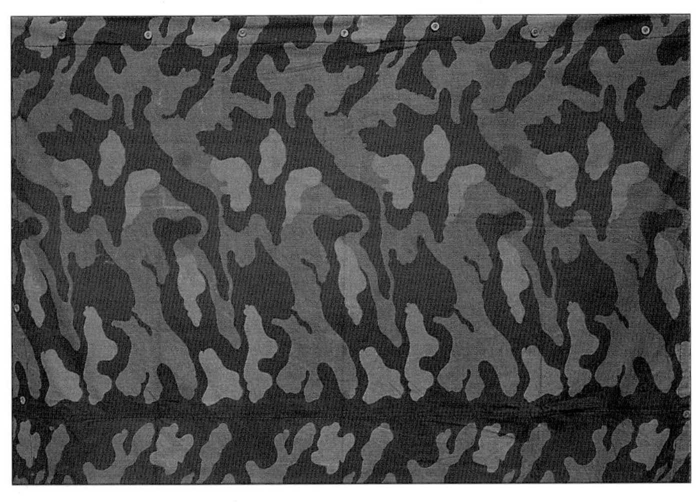

Close up view of 1929 pattern for quarter shelters with dark colors 1939. Markings: SA Bernocchi Colori Chromo America Canapa. Author's collection.

Variant of 1929 pattern for post war camouflage uniforms (quarter shelters, uniforms, and cover helmets). Author's collection.

Archives photo of a fascist jacket 1944 of Salo Republic. 1929 pattern. Private Italian collection.

1950 camouflage uniform for post war, all units except San Marco Battalions. Author's collection.

Special camouflage pattern for "Marines" of Italian navy (San Marco Battalions). Waves are ochre colors on clear blue background. Author's collection.

The new 1980 "Woodland" Italian pattern very similar to U.S. M80 Woodland, for all units (except San Marco Battalions). Author's collection.

Famous elite troops "Bersaglieri" with cock tail pens on the helmet. Y. D. via archives.

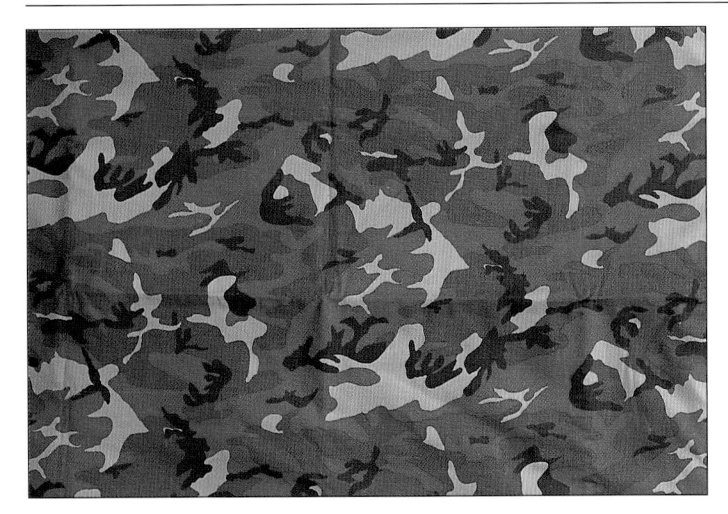

Close up view of "Woodland" new Italian pattern. Author's collection.

Close up view of "pink" pattern of San Marco Battalions. B. Lowry collection, Market Drayton, U.K.

The new "pink" pattern with blurred edges drawings of the San Marco Battalions and with new Kevlar helmet. ECPA Archives, France.

independente." Later the pattern was modified; brown brush strokes became violet, and clear green became dark green, that for 2,000 soldiers of the Portuguese "Marines" and 12,000 soldiers during 14 years in the Angola war.

In 1980-85, a new pattern was adopted, with khaki background, vertical brush strokes and a shorter green. All these patterns were similar to the French "lizard" camouflage suits 1951-1956. A plastic quarter shelter, rainproof, was issued in 1985, yellow background with green and clear brown brush strokes. At present, most of the units of the Portuguese army wear the British DPM 78 (battalion of Portuguese paratroopers), which is also worn by police group GOE (second pattern).

Desert areas new Italian Camouflage pattern. Same drawings as Woodland, but different colors. Courtesy of Italian Embassy in Paris and Head Quarters photo Service in Rome.

Close up view of camouflage pattern for Italian desert area operations. Author's collection.

Spain

Spanish army introduced camouflage uniforms for elite troops in 1952, with two patterns, one green and the other brown, with shapes at fluffy edges. Since 1960, this army has produced many patterns for soldiers, and moreover many different patterns for collective tents—too many to be displayed in the study of a book reserved for camouflage uniforms.

In 1962 there appeared a long series of jigsaw patterns, with green, mustard, gray, brown, pink, background, which would be tested during twenty years in maneuvers (rocks, mountains, meadows, woodland, desert, urban).

Luxembourg soldiers at parade with national badge on the cap. U.S. Woodland pattern. Private collection.

Luxembourg soldiers at shooting training with U.S. M81 Woodland camouflage or a similar camouflage. Courtesy of military Attaché of Luxembourg embassy in Paris.

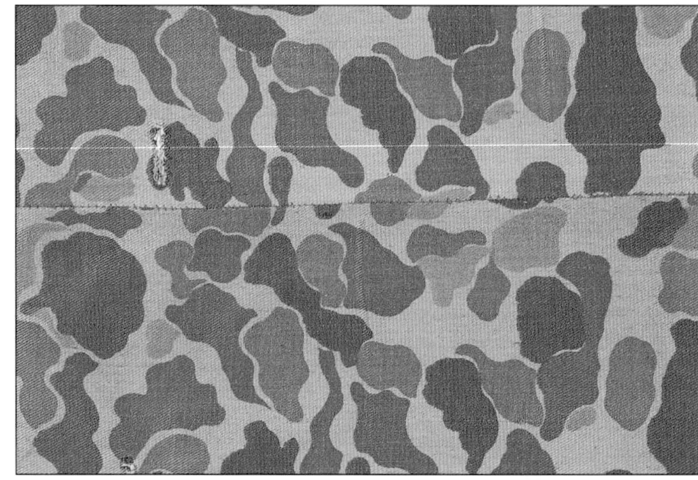

Close up view of Dutch pattern '49-50. Markings: Liebenboom Code 849-51-DLM. Author's collection.

Dutch pattern during independence war of Indonesia 1949-50. This pattern, similar to the U.S. 1944 King Kard, has often been sold by merchants as "USMC Pacific 43 souvenir."

"Marinier" and "jäger" units wear English DPM or U.S. Woodland. No national pattern. Pictured is a captain of the "jäger." Courtesy of SIRPA Terre Archives, France.

A group of Dutch soldiers in British DPM.

Here, a commander of a Dutch "Marinier" unit. Courtesy of NATO photo service, Brussels, Belgium.

First pattern of Portuguese army for Angola wars: light green, dark brown brush stroke on clear white yellow background. Private collection.

Close up view of first Portuguese camouflage pattern. Author's collection.

Second Portuguese camouflage pattern with vertical brush strokes like the French "lizard" 1952-56. Exported in Angola (MPLA), Brazil, and Guinea. Author's collection.

The author himself with second Portuguese pattern.

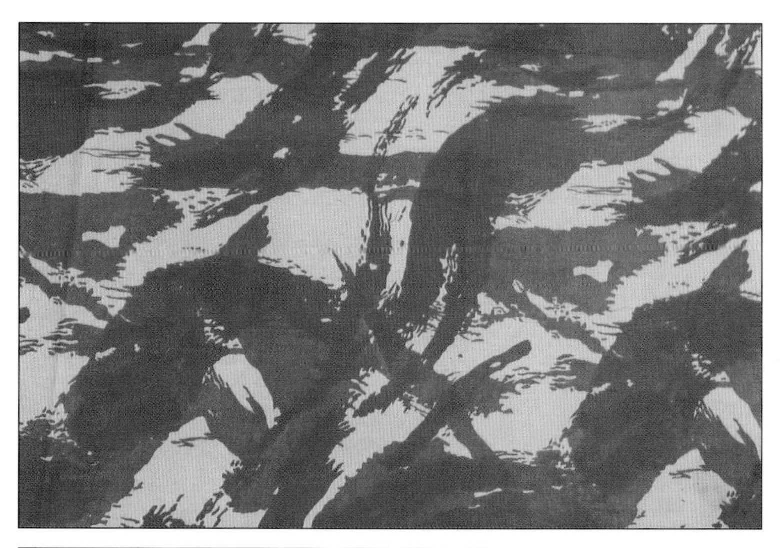

LEFT: Close up view of Portuguese pattern, green vertical brush strokes and oblique violet strokes. Author's collection.

Later in the eighties, the Portuguese pattern was modified: khaki background instead of gray, vertical brush strokes. Author's collection.

RIGHT: 1962 paratroopers' jump suit, autumn pattern. Author's collection.

Portuguese officer at left. Portuguese army presently wears the British DPM camouflage pattern in some units. ECPA Archives, France.

1962 Spanish paratroopers' jump suit summer pattern close up view of material. Author's collection.

1970 Spanish green summer camouflage suit with five colors jig saw. Author's collection.

Close up view of five colors jig saw pattern for two piece suit, bags cover helmets, tents, and ponchos. Author's collection.

A group of Spanish soldiers in 1962-82 camouflage pattern. Private Collection.

The desert areas Spanish pattern for the "legion" in North Africa. Author's collection.

The 1980 "rock mountains" pattern used often by "Marines" and mountain troops. Author's collection.

The "frog skin" material for special operations and for lonely jumps behind enemy lines. Private collection.

Spanish post 83 pattern similar to U.S. M65 67 ERDL. Used for all units, but made in Spain. SIRPA Terre Archives, France.

Close up view of new material, not exactly similar to U.S. army M65 or 67 ERDL. Author's collection.

Two Spanish captains read a map during maneuvers with the French army. They wear the recent pattern 82-83. Private collection.

In 1982 many patterns disappeared and new uniforms were created, such as a special "frog skin" pattern for night secret operational jumps. Very few were worn, and it never was a regulation pattern. A sort of US M65-67 leaf pattern appeared later, but drawings are more or less different. The pink pattern was used again for the "Spanish legion" in North Africa. A snow pattern, with black twigs included in white background, could be seen in exhibition 1988 and Defensa 1989 (Spanish military magazine). In Ceuta, Tercio de Armada wore the desert pattern, and the "leaf pattern" could be seen in maneuvers "Tramontana" in 1994 and Lusitania operations with SGTPAC (tactical airborne troops).

Switzerland

The Swiss army has been greatly influenced by German patterns for one reason: the country in 1940-44 was obliged to manufacture German camouflage material and took advantage of the post war stocks to provide for its own army. So, the Swiss army wore the Wehrmacht patterns after WWII, until 1950 for uniforms, and 1970 for quarter shelters.

In 1955, the famous German leibermuster, made to resist infra red detection, was tested by the Belgian army and the young Federal Republic of Germany. The Swiss army chose this pattern, which is interesting because its own colors—red green, brown, black—seem very appropriate for autumn forest and urban camouflage. This pattern was used until 1997, but reservists used it again in 1999 for maneuvers.

In 1992-95, a new pattern was adopted; the red shapes disappeared, and very few drawings were modified. Many other patterns

RIGHT: First camouflage of Swiss army since 1960. Very similar to German 1945 leibermuster. Red and white, green shapes made for urban fights or in built up areas.

Close up view of Swiss pattern made for uniforms, cover helmets, and tents out of plastic rain proof material; here with a brown background. Courtesy of J.P. Soulier.

Green background for tents, often reversible material for collective tents.

Light summer material of Swiss army in 1978-80. Author's collection.

Reservist soldiers training in the 1960 pattern. Courtesy of J.P. Soulier.

Trial material of Swiss army for tests and peg bags. Never an issued pattern. Courtesy of J.P. Soulier.

The new camouflage pattern of the Swiss army, sort of Woodland, 1986. Courtesy of photo service of Swiss Headquarters in Bern.

Swiss soldiers with the new camouflage pattern. Courtesy of J.P. Soulier.

Close up view of new material. Red shapes have been replaced by brown shapes.

This trial for the Swiss army was never adopted, but drawings were the same as former ones. Courtesy of J.P. Soulier.

"Denison smock" used by British paratroopers from 1943 until 1975.

DPM (disruptive pattern material) of British army 1958, modified in 1978. Y.D. via archives.

have been tested before this choice, but at present, the "Swiss Woodland" has been definitively adopted for all units. This new pattern could be seen recently in maneuvers of the 1st company of pioneers of the 12th battalion (summer 1997, for example).

United Kingdom

After 1945, the British army would use the Denison smock and the light overall, during a rather long time, for troops in India, Burma, Malaysia, until acknowledging the independence of these countries. A large part would be given later to the French army in Indochina and to overseas armies when they became independent. During the Korean war in 1950 the UK army wore wind proof camouflage (29th brigade).

The new camouflage, DPM (disruptive pattern material), would soon replace obsolete material, but the Denison smock (or a similar pattern) would still be used for the paratroopers. These sort of drawings were introduced in 1972 for general use in UK forces, with the pattern confirmed by the PRE (personnel research establishment).

In 1976 a tropical pattern combat uniform was introduced, a special DPM with the same drawings printed in rich shades.[1] During the Gulf war, the army had to use a special pattern tested many years before, made of shapes white, brown, sand, clear brown, but it was impossible to fight with this pattern, which was sold in 1980 to...Iraq!

A new camouflage was rapidly created made of simple brown spots on a clear sand background. It was a very lower quality of

Close up view of British DPM with round shapes, clear green and brown.

Female corporal with the bright colors of modified DPM (United Nations forces). Y.D. via archives.

drawings, but sufficient to protect British soldiers during the Gulf war. An urban camouflage—not military—made in England, and made from the same drawings as the 1972 DPM, has been sold in many police units of Middle East countries, but is not used at present in the British police.

The British DPM green, brown, and black shapes can be seen, for example, in Royal Marines, Antiterrorist Comacchio company (Winged Crusader Operation), leading parachute battalion, parachute battalion group, LPBG, 5th brigade airborne (Pegasus Strike), 1st Royal Gurka battalion, and many other units of the United Kingdom. During the Falklands war all units had received plastic rain proof ponchos and parkas because of the rainy climate, but the colors of DPM were absolutely not appropriated because of the dark brown and black rocks of these islands.

Notes:

[1] See *"British soldiers" in the 20th century, 1980*

Desert areas pattern of the British army. Commonly distributed in Middle East armies, such as Saudi, Koweit, Emirates, Bahrein, Oman, and…Iraq. Author's collection.

Close up view of first pattern "desert areas" with DPM drawings, but in a variant of brown and sand colors.

Close up view of new desert camouflage distributed just before the Gulf war against Iraq.

Special camouflage for the Gulf war, simplified and different from the obsolete desert camouflage of the seventies sold to Iraq.

The urban camouflage, made of variants of blue, is mainly used in African police units and is rarely used in the United Kingdom. Private collection of a U.S. collector, Springfield.

3

Camouflage Uniforms of Northern Europe

Denmark

The Danish army seems to have adopted a camouflage suit in the seventies when Texunion France factories proposed a pattern to this country. Only one camouflage type has been chosen and was distributed to all units. Green with black, on a yellow white background, this pattern has a very good camouflage efficacy in woodland.

Estonia

The Estonian army wears at present time a green and dark brown camouflage pattern provided by the Russian army, since 1980. A special sand pattern has been created, maybe for the large beaches of this country. The drawings of the sand pattern are the same as the woodland and are made of yellow, pink and sand shapes.

Female Danish officers of the royal army wearing the camouflage pattern 76-80. Courtesy of SCPO K.L. Jeppsson information branch defense command, Denmark.

Close up view of Danish pattern, the same as German army but without brown dots and shapes.

Danish soldiers in heavy padded camouflage parkas and flak jackets during mortar training.

A group of Estonian soldiers with three blue variant color lions on the shoulder patch. Y.D. via archives.

LEFT: Danish NBC protection one piece coverall. Note the excellent camouflage effect in Woodland. Henrik Clausen photo.

Estonian soldiers with Navy camouflage pattern of Russia type II. Y.D. via archives.

National Finnish camouflage in 1970. Courtesy of Mr. Paassikivi, Hamenlina, Finland.

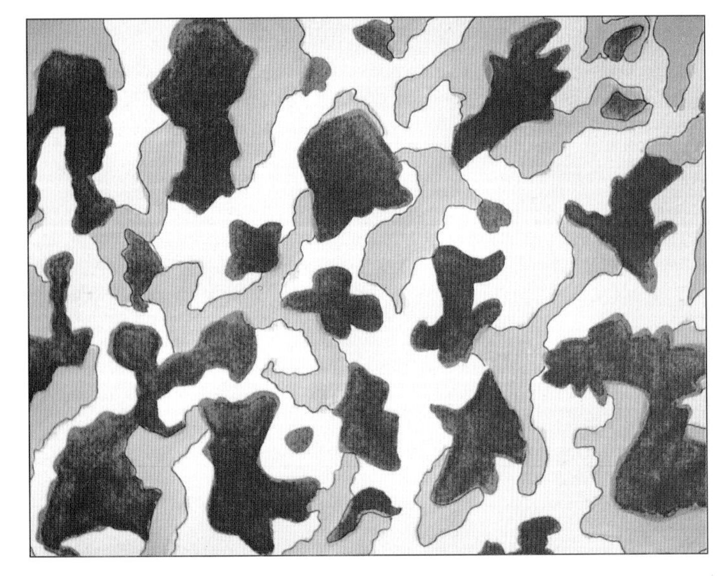

Special Estonian camouflage pattern for sandy beaches with three variants of brown. Author's drawing.

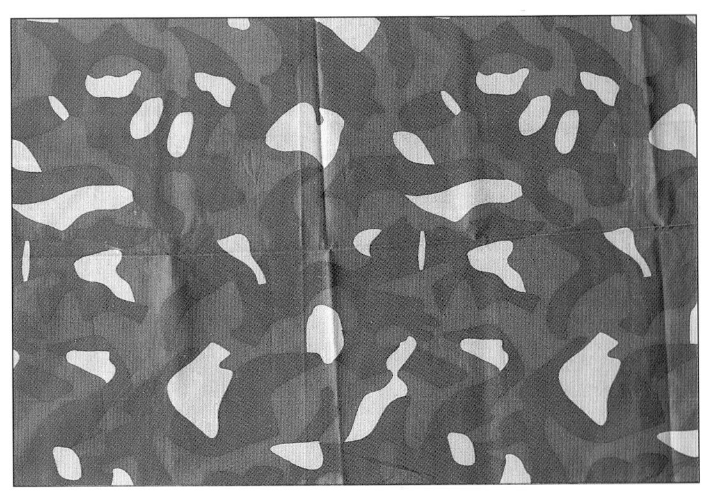

Close up view of Finnish material for camouflage 1970. Author's collection.

Finnish variant of colors and also special light material to cover the heavy padded material in winter: clear green-gray.

Finnish variant of colors tested in 1980 but never adopted. Author's collection.

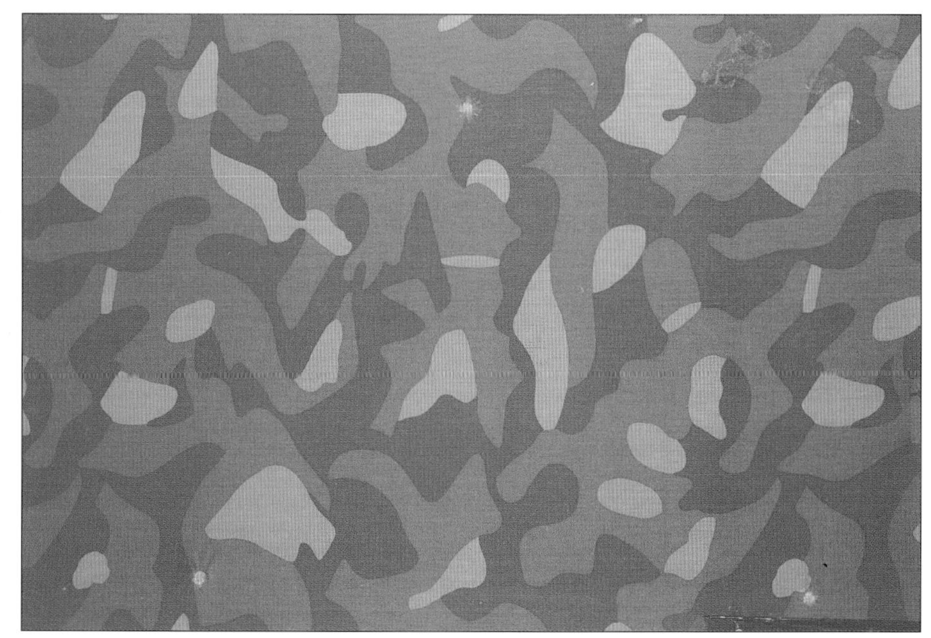

Close up view of khaki background pattern for winter Finnish material. Author's collection.

Finland

Since 1970 the Finnish army has worn a camouflage uniform which has special colors tested for many landscapes. So, four or five patterns have been worn during many years, always with the same drawings, middle size shapes generally made of clear colors and different backgrounds: khaki, dark green, and clear green. Recently a pattern has been adopted—the 1990 pattern—with apple green and ochre shapes.

Other variant colors for winter padded material and parkas. Never adopted.

The M 90 Finnish pattern definitively adopted for all units. Courtesy of Julkaisuoikeus Puolustusvoimat Pv KK Kuvaosasto. Käytettävää Tekstia SA Kuva.

Iceland

This country has neither an army nor a camouflage pattern.

Latvia

The Latvian army wore during its communist period a special pattern for this country.[1] The pattern was made of sorts of "tadpoles" brown on gray or khaki background. At this time there also existed an urban camouflage pattern made of large different blue spots and shapes.

In August 1998, a curious woodland pattern could be seen in Riga, similar to US M65-67 ERDL of the Vietnam war, but with a clear green background. At the present time, the Latvian army wears a sort of U.S. Woodland, and during maneuvers with NATO exercises, this army wore the U.S. M81 pattern.

Lithuania

The Lithuanian army, after its communist period, has created a national camouflage pattern, gray background and many little spots, long and thin, green and brown. Since maneuvers with NATO in the U.S. and Europe, the Lithuanian army seems to have adopted, for certain units, the U.S. M81 Woodland. But the national pattern is worn again at present.

Norway

The first camouflage pattern of the Norwegian army was made for air force commandos to survey military airports. Dark brown shapes with superimposed colors and drawings, it was soon dropped in 1970. A new pattern replaced it, made of long brown and green waves on khaki background, worn by all units. For summer time, special light material has been issued, with the same long waves, but dark green and black, and no brown waves. A special camouflage is used for collective tents.

Sweden

The Barracuda factory in Lahölm was in the sixties the big establishment oriented to camouflage material. A couple of colors had

Soldiers wearing material with pink ochre background and apple green or dark green shapes (Finnish maneuvers).

Latvian camouflage pattern made by Russian army, 1993. Author's collection.

been tested, with a very modern concept: large black splinters and white, green, and pink circles. A camouflage pattern of this style had been created, but also many patterns for vehicles. The ACMAT society in Saint Nazaire, France, made military vehicles with this curious camouflage, but it soon disappeared. Two patterns existed, woodland and desert areas.

Later in 1985, Norsel Fabrik AB created a "splinters" pattern dark and clear green, white shapes, worn by all units, but also vehicles. The old quarter shelter with blue and gray long waves has been conserved.

Notes:

[1] See Desmond, Dennis. *Camouflage Uniforms of the Soviet Union and Russia, 1937-to the Present,* p.123.

A later Latvian camouflage pattern made in Russia (see D. Desmond book, p.123).

Blue urban camouflage of Latvian army made in Latvia. Bruno Renoult collection.

LEFT: Latvian camouflage seen in Riga by the author in August 1998 (maybe police forces?).

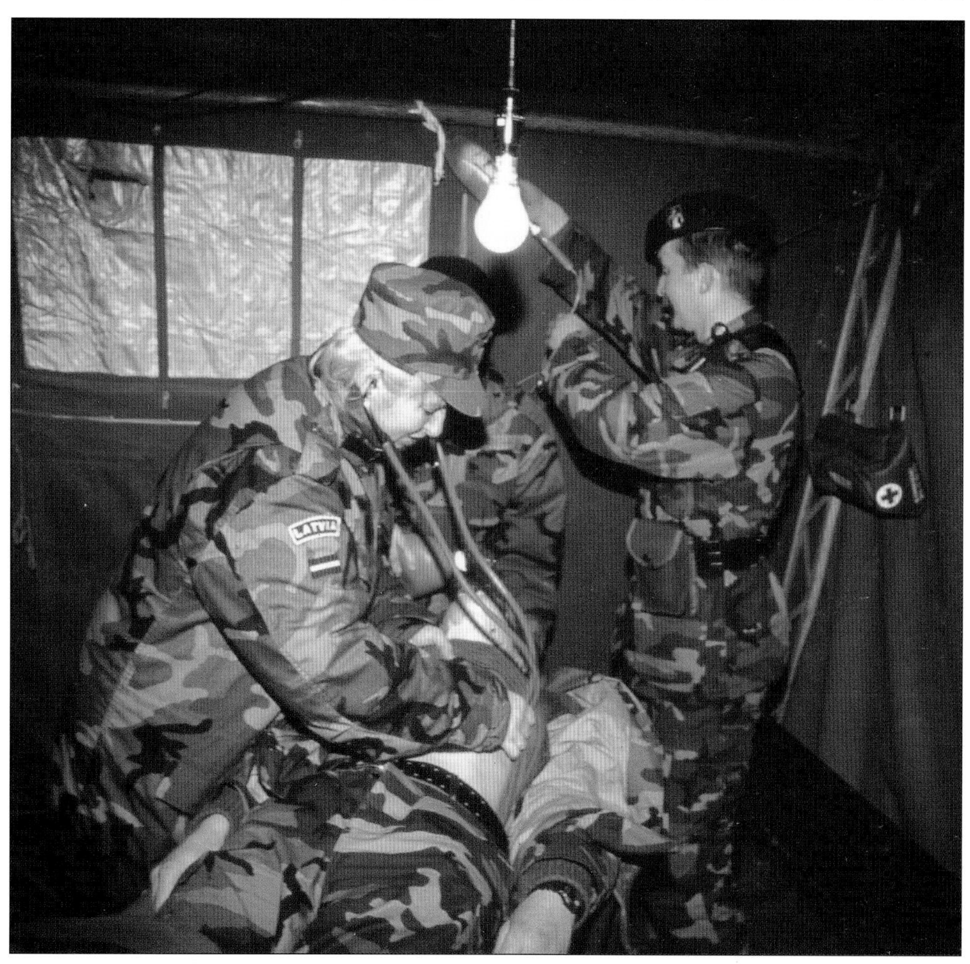

A group of Latvian doctors during maneuvers wearing a U.S. camouflage suit.

LEFT: The new camouflage pattern similar to U.S. M81 Woodland. Courtesy of Ministry of Defense in Riga 1998.

1997 Lithuanian camouflage pattern. Will soon be replaced by U.S. M81 Woodland or similar.

The new Lithuanian camouflage, which is very similar (or actually is) U.S. M81 Woodland. Courtesy of Krasto Apsaugos Ministerija. Fotografas Tadas Dambrauskas.

Lithuanian soldiers at parade with the national flag. Courtesy of Krasto Apsaugos Ministerija.

Close up view of camouflage pattern with green and brown shapes, on pink gray background.

Rare camouflage pattern, soon dropped in 1975, of air forces Norwegian commandos.

Norwegian troops wearing heavy padded parkas in winter. SIRPA Terre Archives, Paris.

The new Norwegian camouflage, long waves green and brown on a khaki background. Private collection.

The long winter parka camouflage of the Norwegian army. Author's collection.

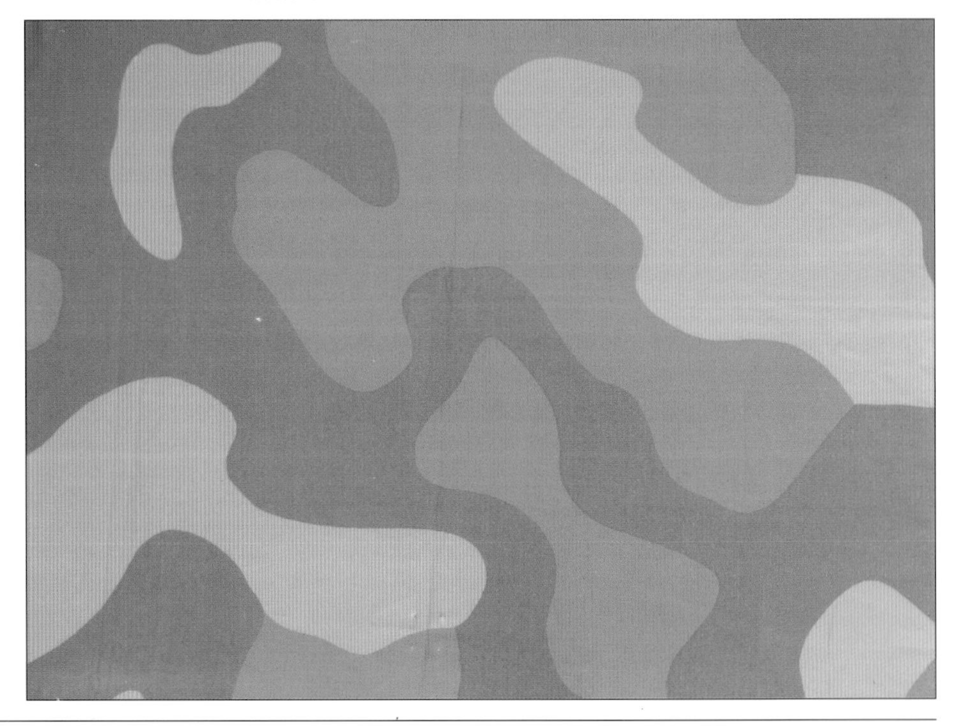

Closer view of Norwegian camouflage pattern. Author's collection.

Summer camouflage uniform (without brown waves) for Royal Guard's soldiers (September '98. Exercise cooperative "Best effort" in former Yugoslavian Republic of Macedonia. Krivolak training area). NATO photo service, Brussels, Belgium.

The Swedish Barracuda pattern (Lahölm manufactures, 1975). Soon dropped.

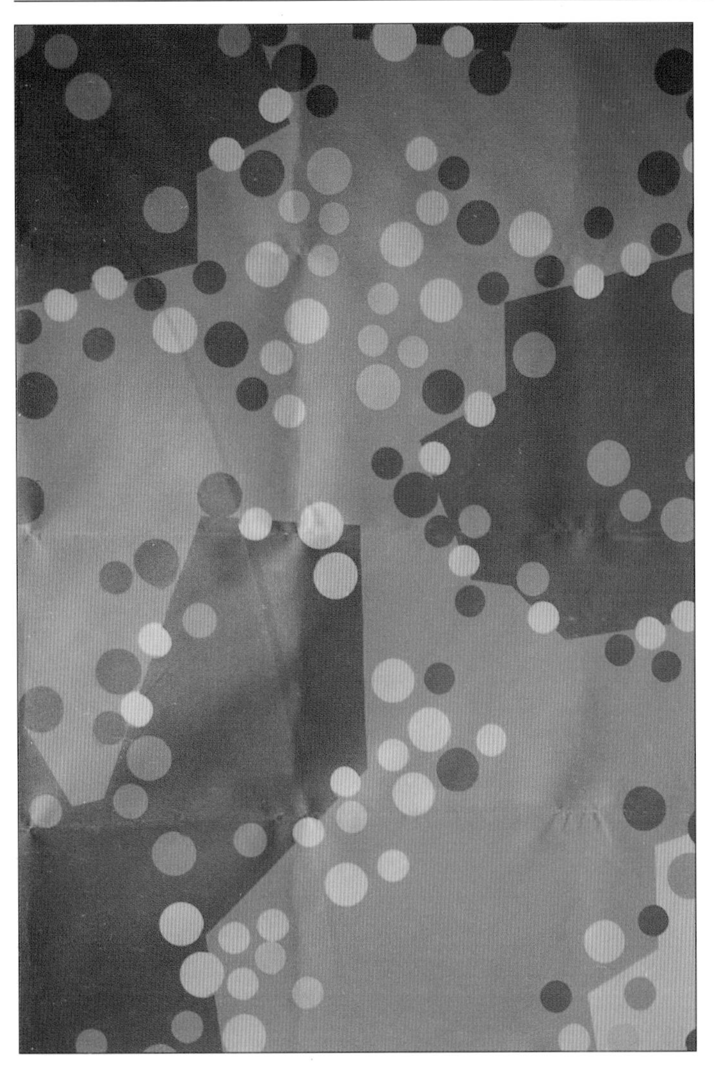

Close up view of Swedish Barracuda pattern with black splinters and pink/black/green regular circles. Author's collection.

New 1980 Swedish camouflage pattern "splinters" for personnel and vehicles. (Norsel Fabrik AB). Author's collection.

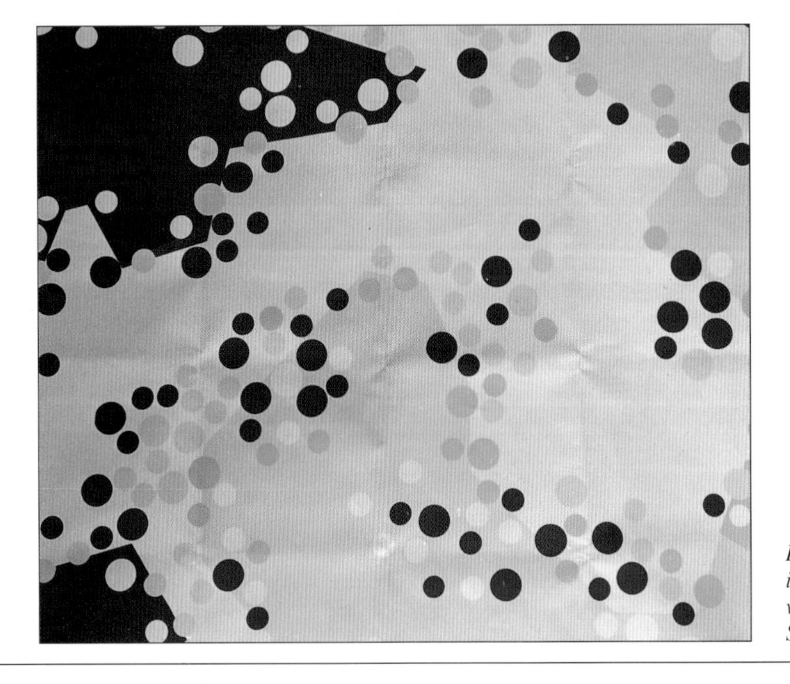

Desert areas pattern from Swedish Barracuda for tarpaulins and vehicle covers. (Vehicles ACMAT, Saint Nazaire, France).

Swedish soldiers at training with two piece camouflage "splinters."

Close up view of Swedish splinters camouflage pattern. Author's collection.

Poncho tent gray blue brown used again presently in the Swedish army.

4

Camouflage Uniforms of Central Europe

Albania

Just after WWII, the Albanian army received during the "cold war" many camouflage patterns TTS MKK from the USSR, but mainly the KLMK 68-71 since 1980 when old camouflage was faded. Recently the Albanian army has participated, along with the U.S. army, in the Exercise Cooperative maneuvers, assembly 17-22 August 1998, and now wear the U.S. M81 Woodland.

Bosnia Herzegovina (Republic of Bosnia Herzegovina)

All along in the post-Yugoslavia wars and during many years, men at arms in this country have worn many different camouflage suits. It is very difficult at present time to say what pattern has been chosen. Only one appears to be a sort of national pattern, with a background sprinkled with little points, and sharp pointed drawings ochre red, brown and green. But many Moslem units also wear a sort of U.S. M81 Woodland, maybe made in Bosnia.

Albanian soldiers at parade wearing a soviet camouflage 1950 TTs MKK pattern. Ushtria Jone Popullore photo, Shtepia Botuese, Tiranë.

Close up view of soviet trial 1945-50 distributed to Albania. Author's collection.

1996 USA M 81 Woodland for Albanian troops (Exercise Best Effort. Camp Lejeune, USA, 23-27 August 1996). NATO photo service, Brussels, Belgium.

Albanian soldiers in Exercise Cooperative assembly at Rinas airport in Albania, August 1998. USA M 81 pattern or similar. NATO photo service, Brussels, Belgium.

Bosnian camouflage in 1997, worn only by some units. Author's collection.

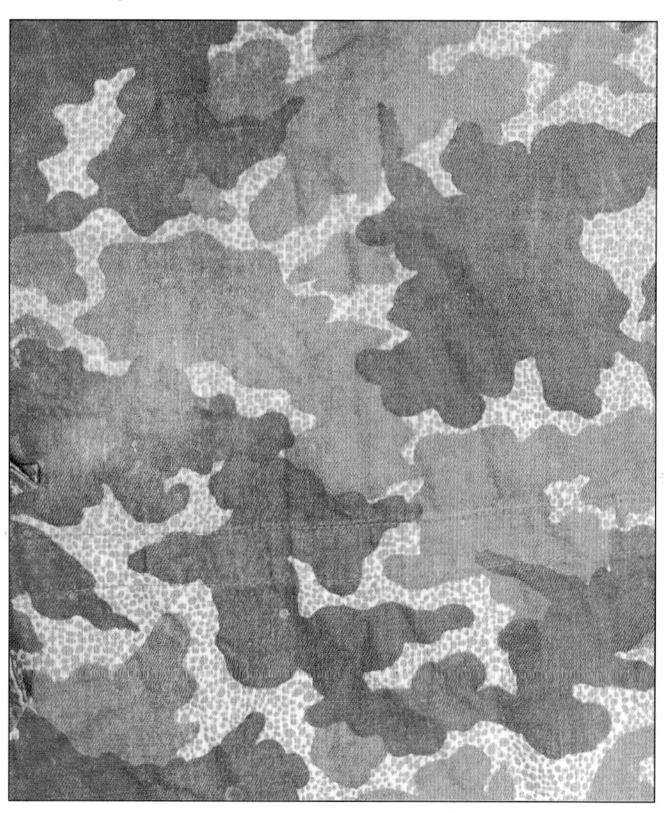

Close up view of Bosnian pattern. Author's collection. Note that Croatia-Bosnia confederation wear a similar U.S. M81 pattern, and on the sleeve a badge with Europe 12 stars, lis flower and the Croatian white red checker board.

Bosnian soldier observing the front. Y.D. via archives.

Croatian soldiers wearing a camouflage pattern similar to the U.S. M81 pattern in south Croatia.

Croatia (Republic of Croatia)

Exactly as Bosnian Troops, many Croatian soldiers have worn during post-Yugoslavian wars many different camouflage suits. But since independence of this country, it seems that all units wear, here again, similar M65 or M81 U.S. army camouflage patterns. But again can be seen the old ground shelter apple green with long yellow and brown twigs.

Czech (Czech Republic)

Many patterns have been tested since 1950, because Czech military authorities refused to wear any USSR pattern. The first pattern was a quarter shelter made of indescribable shapes with blurred edges, fluffy spots, which was also tested as a uniform, but soon dropped; later appeared as so paradoxical a pattern, and so high colored (green, yellow, orange) that it could not really be considered a camouflage pattern. But this uniform had a large efficacy in town, among colored streets of built areas.

In 1955 the "clouds patterns" invaded the Czech army, white, green, ochre, brown clouds, and a sort of "leibermuster" like the famous German pattern of February 1945. Later, the Czech army was influenced by a recent German pattern, at a moment where

Croatian camouflage in 1998, shirt with similar U.S. M65-67 ERDL. Trousers probably made of another material and similar to an Italian pattern. Y.D. via archives.

Close up view of Rip Stop similar to U.S. M81 camouflage. Badge of Croatian SJP Zagreb unit. B. Lowry collection.

The 1950 quarter shelter of the Czechoslovakian army of the Warsaw pact. Author's collection.

every army of the Warsaw Pact wanted to wear the same uniform, gray background and rain strokes. Czech, then Czechoslovakia, chose such a pattern, with few green shapes visible in the background.

It would be the last camouflage uniform before the fall of the Berlin Wall and the liberation of ancient "allies" of the USSR. In 1993, the Czech army separated from Slovakia, selected a very new camouflage pattern, green background and long yellow and black sorts of stars, which was worn in 1997. This recent uniform could

LEFT: First "clouds" Czech pattern white gray brown colors, with "beaver tail" of paratroopers. Faded material. Warsaw pact. Dan Peterson collection.

Second "clouds" Czech pattern brown turquoise gray colors. B. Lowry collection.

RIGHT: Czech urban camouflage, very brightly colored, perhaps for fighting in built up areas, 1975. Author's collection.

New Czech rain strokes and blue faded drawings in the background. Y.D. via archives.

LEFT: Third Czech "clouds" pattern green black white, 1970. RIGHT: Czechoslovakian similar to German "leibermuster" for fights in built areas. Reversible with the "clouds" pattern pictured at LEFT. Author's çollection.

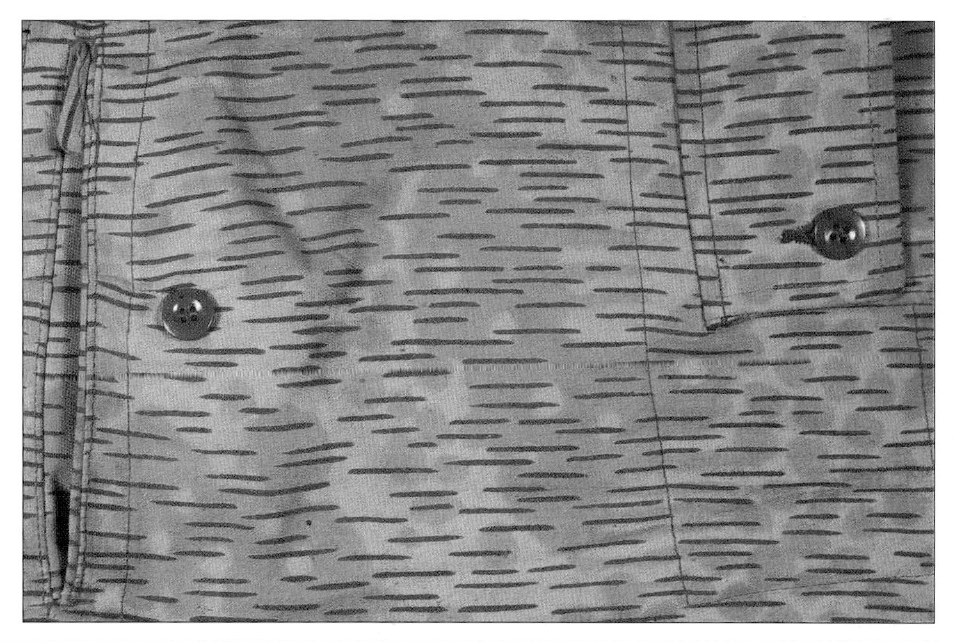

Close up view of Czech with reversible drawings, blue among the rain strokes.

Rain strokes on camouflage jacket, close up view. Author's collection.

be seen already in 1993 at Boletice in Bohemia training camp, and also during maneuvers with the 4th intervention brigade.

Greece

The Greek army has selected since 1965 three sorts of camouflage patterns. Always brush strokes, but in various positions—horizontal, diagonal or vertical. These three patterns have been tested during twenty years, and horizontal brush strokes on an olive drab background has been retained.

The first test of a green background with vertical brown and dark blue brush strokes was tested on "Marines" of the Navy. The second test was made with a gray background, brush strokes dark green and brown, but was very soon dropped.

The paratroopers in the summer of 1980 wore a T-shirt with the badge of Dynameis Katadromon, with different green "frog skin"

The new camouflage of the Czech army after the separation from Slovakia. Y.D. via archives.

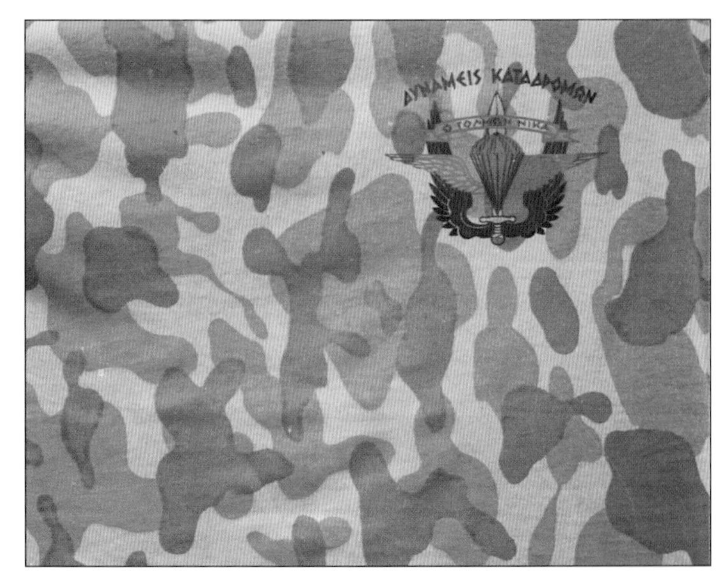

Greek T-shirt with frog skin camouflage and the badge of paratroopers (Dynameis Katadromon) for summer and fatigue garments.

Close up view of Czech camouflage. Courtesy of Oldrich Jerabek, Vasiscova 505, 27204 Kladno 4.

First dark green trial for "Marines" of the Greek Navy (Dynameis Pezonauton) black and green vertical brush strokes. Author's collection.

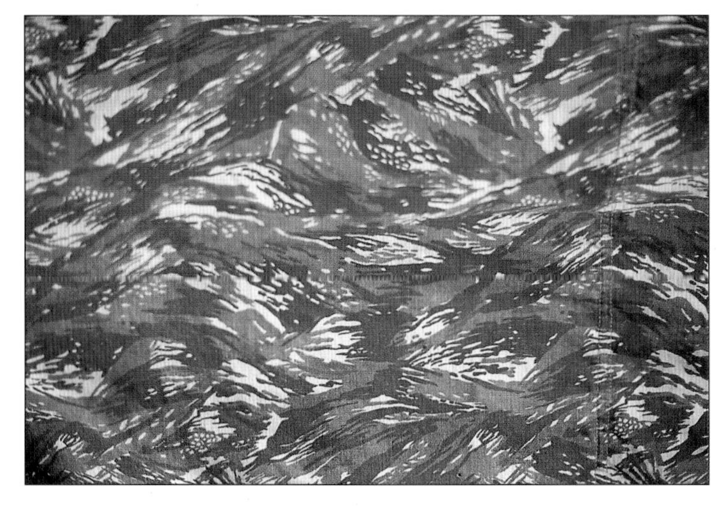

Close up view of material in trial for Greek "Marines."

Close up view of second Greek pattern soon dropped. Author's collection.

Second pattern trial for all units, but not worn often in 1976.

Definitive pattern adopted by the Greek army and similar to French lizard 1952-54. The background is olive drab; brown and green brush strokes.

Soldier of a Greek unit with new camouflage already tested in 1975 and finally adopted later for all units. The soldier tests a Georgian recruit in shooting (See the Georgian camouflage suit). Y.D. via archives.

1945 Hungarian camouflage shelters at the end of the Russian campaign. ECPA, Archives, France.

shapes. At present, the Greek army wears only one camouflage pattern, very similar to the French "lizard" brush strokes.

Hungary (Republic of Hungary)

During the period 1939/40, Hungary, Germany and Italy were the only countries to introduce a camouflage material for the protection of their Axis forces. Neither Romania or Bulgaria, nor Slovakia or Bohemia Moravia introduced camouflage at this time, as did the other German allies.

Hungarian patterns were divided into both summer and autumn patterns used on large half tents able to be worn as a poncho. Designs were similar to the Italian type for the summer material,

The two quarter shelters of the Hungarian army at the end of WWII. In the center is the Italian quarter shelter. J. de Fromont and J. Borsarello photo.

Green coverall one piece of Hungarian snipers during the Warsaw pact. Author's collection.

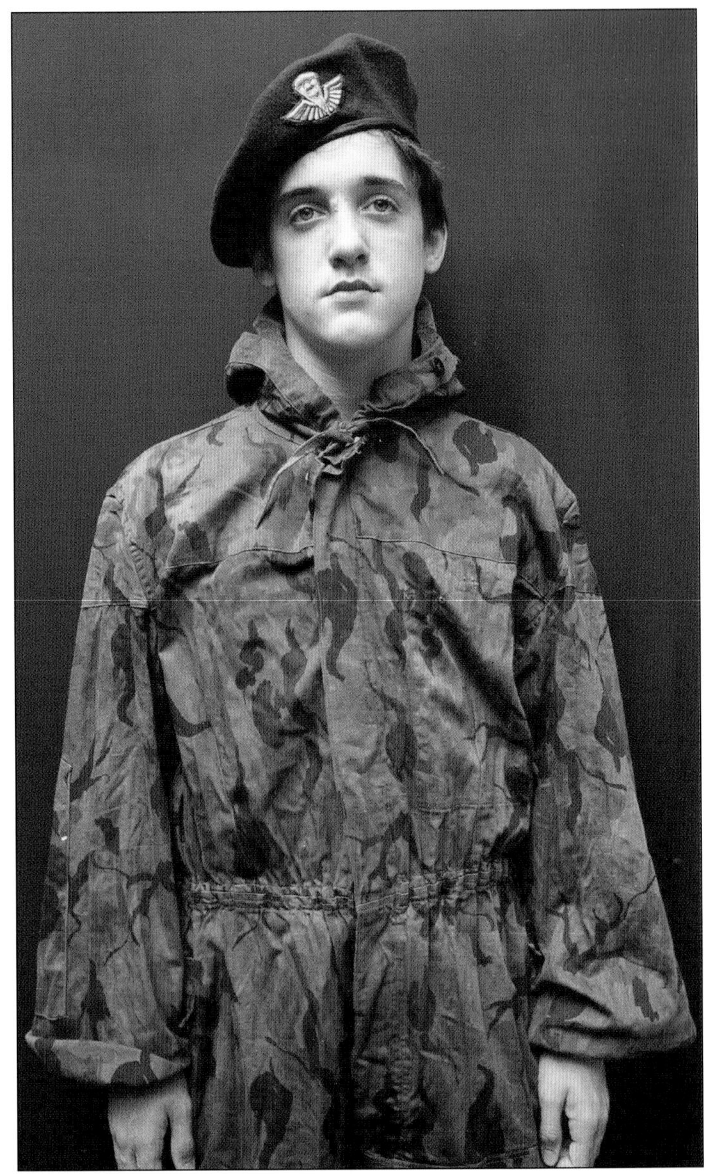

Hungarian khaki background and brown drawings coverall. Courtesy of Hadtortenety, military museum of Budapest.

but were very different for the autumn material, pink bowls and similar to musical notes on a dark brown background.

In the communist period, 1950-1980, four patterns existed, worn during the same times as trials: a one piece, overall garment, with hood and face mask was generally made of clear green or dark brown background, sprinkled with little twigs or branches, reserved to snipers. The second pattern was very colored, made of yellow, green and brown leaves with tooth saw edges, reserved to general purpose, in meadows or green landscapes.

The last pattern was similar to the 1943 half tent, brown background, pink bowls, sorts of "musical notes" and green "fingers" superimposed, blurred edges shapes. It was a two piece suit or paratrooper's uniform seen during the peace keeping campaign, or Cooperative Lantern 1998 in maneuvers Balaton Kenese, for example.

Macedonia (Republic of Macedonia)

The young Macedonian army has rapidly selected a very special national camouflage pattern, with more or less horizontal stripes, green khaki and brown, very thin and upon a clear yellow background. This pattern has been tested in maneuvers with the Georgian army in Skoplje (first battalion "Scorpions" of 11th motorized brigade: Cooperative effort exercise).

Slovakia (Republic of Slovakia)

Just after independence, Slovakian forces of UNO in post-Yugoslavia wore a sort of "brown woodland" pattern, which could also be seen in the booklet published by the new Slovakian army.

But certain units wear, at present time in 1999, the Czech army 1993 pattern. This new pattern could be seen along with the old pattern in 5 pluk Specialnelio Urcenia and Rota Specialnelio

Urban camouflage two piece suit four colors pattern of the Hungarian army (similar to DDR 1970 pattern with ragged edges leaves). Author's collection.

1988 new Hungarian camouflage uniform with same pattern as 1940 brown-pink balls quarter shelter. Here a paratrooper captain.

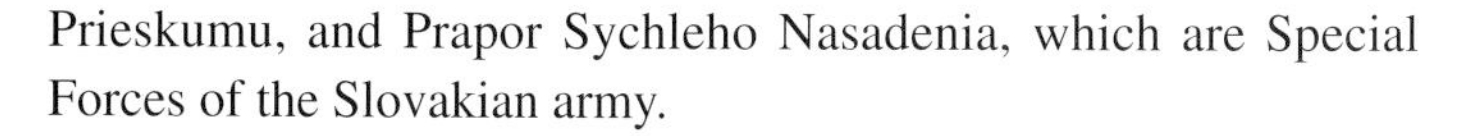

Prieskumu, and Prapor Sychleho Nasadenia, which are Special Forces of the Slovakian army.

Slovenia (Republic of Slovenia)

The Slovenian army has worn during some years the same similar U.S. M81 Woodland as the Croatian army. The Specialna Brigada Moris, at present time, wear again this pattern, possibly really a U.S. uniform.

After a short time, it could be seen in a booklet published by the Slovenian army that a new camouflage had been selected. (Territorialna Obramba Republike Slovenije). This pattern is made of superimposed shapes, little areas, brown, green or black. The background is gray colored.

Hungarian officer with recent camouflage pattern, pink balls as music notes and brown shapes, in 1998. Y.D. via archives.

Recent new camouflage uniform of young Macedonian army: little white spots among khaki and dark green shapes superimposed. Y.D. via archives.

Close up view of Macedonian army pattern.

Yugoslavia (Federative Republic of Yugoslavia)

Under Tito government, the Yugoslavian army has not received any camouflage uniforms and had selected a pattern only in 1970: a half tent and a two piece suit with hood, apple green with long yellow and brown branches without any leaves. This pattern was seen everywhere in civilian activities: tarpaulins for trucks, curtains for stores, and even guitar bags! For mountains and snipers, the Yugoslavian army had created a light coverall with hood and gloves, mask face. Colors were brown pink, and very little invisible points in the background, which displayed here and there fluffy splinters.

Brown background Woodland of Slovakian army pattern in UNO forces in former Yugoslavia. Peucelle photo.

The Slovakian army has also worn the new Czech pattern (booklet: partnership for stability and security, page 3, photo Milan Cipka). During exercises with NATO in 1995 Slovakian forces wore U.S. M81 Woodland.

Close up view of Slovenian camouflage and national patches. Courtesy of Lubljana ministry of Defense and Military Attaché of Slovenian embassy.

Slovenian soldier with the new camouflage pattern. Y.D. via archives.

ABOVE: First camouflage uniform of Yugoslavian soldiers during the Tito period. Also worn during the independence war of former Yugoslavian states. Y.D. via archives.

LEFT: Close up view of the first pattern of the Yugoslavian army 1970-1980. Apple green background and brown or yellow branches without leaves for quarter shelter. Author's collection.

Mountain-rocks pattern for Yugoslavian mountain troops, pink brown and splinters sprinkled of little points for snipers. Author's collection.

LEFT: Jacket made of quarter shelter 1970 of the Yugoslavian army.

Close up view of Cuban-French similar material, gray background for Yugoslavian police.

Serb soldier in patrol with Serb pattern two piece suit. Y.D. via archives.

LEFT: A colonel of Serb militia with "lizard" pattern similar to the Cuban pattern and also the French pattern 1954.

At last, in the seventies appeared a sort of "tiger stripes" pattern, which was soon dropped. Militia wore at this time the "lizard" pattern, very similar to French camouflage, worn also by Cuban troops in Cuba and mainly during the Angola war.

During the post-Yugoslavia wars, many patterns have appeared, but one seems to be very special for Serb soldiers: clear green background with large shapes brown, black and green. The "lizard" pattern, Cuban type, is also worn by Serb troops.

RIGHT: Typical Serbian camouflage pattern 1997 with large shapes brown and green. Y.D. via archives.

5

Camouflage Uniforms of Eastern Europe

Belarus army wears a Russian pattern at the beginning of 1999.

Belarus (Republic of Belarus)
In 1998 the Belarus army again wore a Russian pattern, and it could be seen in photos in a magazine. We never had a reply to our claim for information from this country.

Bulgaria
During the Warsaw pact, two camouflage uniforms were introduced in the Bulgarian army, with rain strokes, splinters, and large shapes with holes. Only colors differ from the German 39-45 army camouflage. In 1998, with a new government, these patterns have been maintained because of their excellent efficacy

The first camouflage issue was made of large areas with holes, and many big points, around, exactly the same as the "plane tree" pattern of German camouflage 1935-45. But here the large shapes are green instead of black. Rain strokes are brown. The second pattern is made of splinters green and brown, like the luftwaffe German paratroopers 39-45. It also has rain strokes.

After the end of the Warsaw pact, the Bulgarian army wore for a short time the Russian "tigerstripes."[1]

Use of camouflage German patterns 39-45 is not a derogatory act if it is considered that camouflage is made, in principle, to avoid injuries and casualties in war. Many countries, such as Libya, Syria, Egypt, Sweden, Czechoslovakia, and Poland have also used obsolete rain strokes, splinters or plane trees.

Cyprus
The Cyprus army generally wears a Greek pattern, but recently, "frog men" commandos have adopted two camouflage uniforms. The divers wear a special diving suit, rubberized and colored with a pattern generally called "Duck Hunters," because these drawings are used everywhere in the world for hunters. It is a similar pattern to the famous U.S. HBT King Kard dress of "Marines" in the Pacific war 42-45.

The second pattern is exactly the old Singapore forces camouflage uniform, worn by this independent country in the sixties. Note

First postwar Bulgarian pattern of the army, 1960. Author's collection.

Bulgarian similar German rain strokes, but waved, and dark green and brown splinters pattern, gray background. Author's collection.

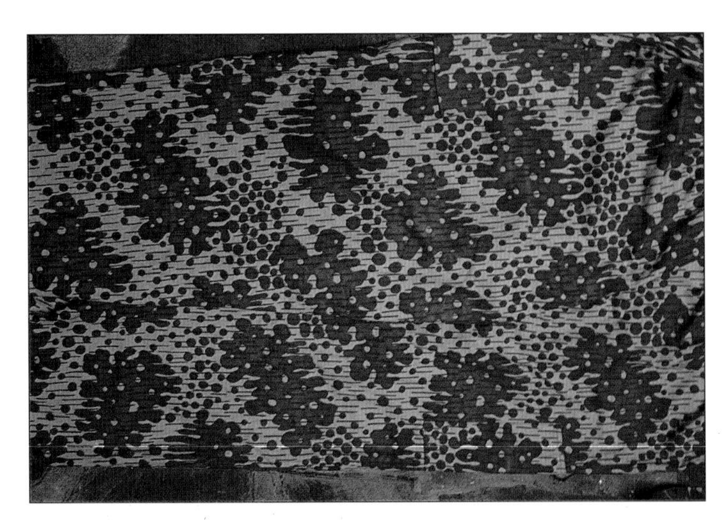

Close up view of 1960 Bulgarian pattern made of large green shapes with holes, green points, and rain strokes. Soon dropped. Author's collection.

that the north of Cyprus is occupied by the Turkish army, and the soldiers, of course, wear the camouflage uniform of Turkey forces.

Moldavia

This young army has selected a similar U.S. M65 or 67 leaf pattern, but wear again, on helmet covers, bags, and accessories the Russian camouflage especially made for Moldavia.[2]

Poland (Republic of Poland)

The first camouflage of the Polish army was a very high colored brown, orange, green, yellow pattern, only available in urban camouflage, or meadows at summer. Soon dropped, this pattern has been quickly replaced by a pattern greatly influenced by the German army 39-45: large splinters green and brown, or green and violet, with rain strokes, very often reserved for paratroopers.

1996 camouflage pattern, splinters and green background. SIRPA Terre Archives, France.

A sort of "tiger stripes" made in Russia seems to have been worn by some units in Bulgaria. Private collection.

Close up view of splinters pattern of Bulgarian army, 1970-1992.

Camouflage pattern for Cypriot commandos. Y.D. via archives.

Close up view of Cypriot pattern for commandos (obsolete Singapore 1970 pattern). Author's collection.

"Duck hunter" camouflage for diving suits in the Cypriot commando. Y.D. via archives.

Camouflage 1999 of Moldavian army, similar to U.S. M81 pattern. On the helmet, Russian pattern made for Moldavia in 1995. Y.D. via archives.

Close up view of colored petals Polish 1960 pattern.

In 1960, the trial of a "panther skin," gray background, was soon dropped. In 1970, there appeared a "camouflage" pattern made of little rain strokes, so many squeezed together that it was impossible to see them at a distance. This uniform seemed gray only. In 1980, appeared the "crocodile" pattern, not very efficient, and even a rain stroke pattern exactly like the German DDR pattern. And again, a last rain strokes gathered three by three!

After the fall of the Warsaw pact, the Polish army has maintained in rare units of "Marines" the "crocodile" pattern, but two different camouflage patterns have been introduced. A plastic waterproof uniform, made of black and orange spots on clear green background is issued to commando troops. The more recent camouflage at last is a turn back to the first one, petals, fingers, brown or green on an olive drab background. But the petals are very little, a reduction of sixty per cent from the original pattern.

Initial patterns of the Polish army in 1960 in the Warsaw pact. At left yellow, orange green, brown large petals. At right yellow background for similar splinters green and brown used in German army 1938-45. Author's collection.

Close up view of green splinters, different brown and rain strokes. Author's collection.

Clear gray with green and violet splinters, rain strokes pattern for Polish paratroopers in the Warsaw pact. Courtesy of Dan Peterson.

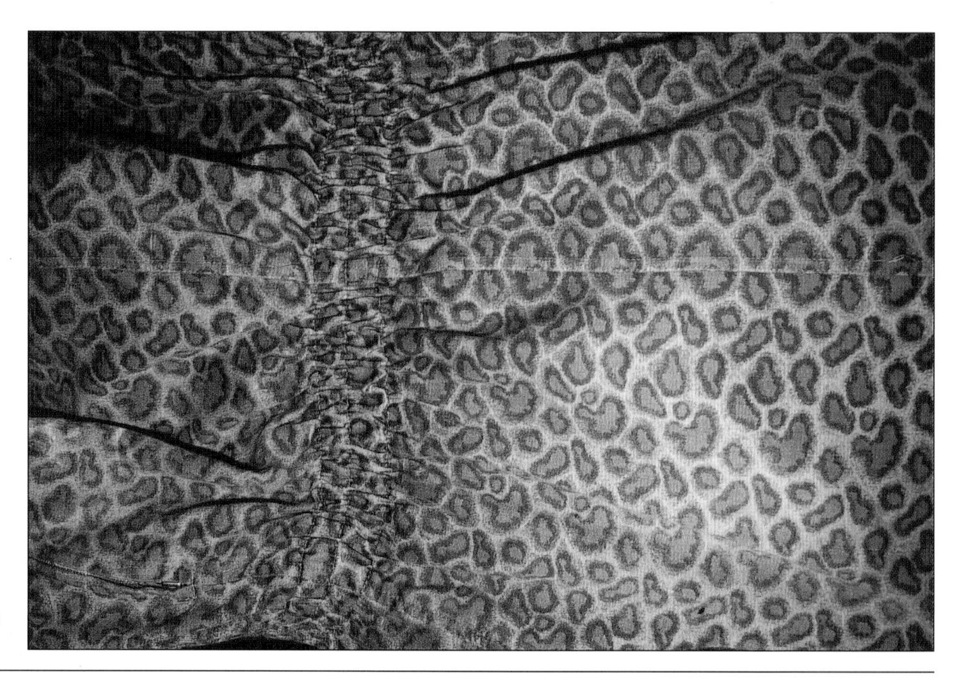

Panther pattern of Polish paratroopers 1975. Soon dropped. J.P. Gillet collection.

Polish commandos in "frog skin" pattern in an exercise, 1997. Y.D. via archives.

Gray background and tiny dashes of 1980 Polish pattern in the Warsaw pact (the armored badge is in the wrong place, generally on upper sleeve). Author's collection.

"Crocodil" dark gray spots for "Marines" in Poland in 1997. This pattern was also used in 1988. SIRPA Terre Archives, France.

Close up view of Polish waterproof plastic camouflage "frog skin" pattern for commandos. Author's collection.

1998 pattern for "Marines" and generally used in Polish army. Petals are similar to the 1960 pattern, but smaller and on a khaki background. Y.D. via archives.

The "crocodile" pattern could be seen during maneuvers of the 11th armored cavalry division. The last "petals" pattern has been used in Polish intervention troops of "IFOR," 16th battalion (Powiertrzno Desantowy) in Cooperative Bridge Operation in Teslic, post-Yugoslavia 1994. The "frog pattern" plastic was worn by the 6th division, assault brigade "Sosbowski" and "reconnaissance unit GRN (Grupa Rozfoznawco Naprovadzajaca)."

Romania

During the communist period, the Romanian army was supplied with Russian camouflaged uniforms: small white leaves on a green background (summer), and small white leaves on a brown background (autumn). A parka with hood and trousers designed to be worn over the standard uniform was also created. In 1985-88 a simi-

RIGHT: A major and a captain of mountain troops with 1998 pattern little petals. Courtesy of Wojskowa agencia fotograficzna. Stawomir Kaczorak photo. Prawa autorskie zastrzezone.

USSR pattern for Romanian army in 1960 in the Warsaw pact; little white leaves on green or brown background. Author's collection.

Close up view of brown side "autumn" of Romanian coverall 1960.

Brown background of USSR camouflage pattern for Romania. Author's collection.

"Persilla" pattern of Romanian army worn as coverall and hood, similar to USSR design for Warsaw pact patterns. Author's collection.

After the fall of the Warsaw pact, a new Romanian camouflage emerged. Here is a medical doctor in blue cap and a male nurse observing a French soldier.

lar uniform was introduced, but the white leaves were replaced with small green parsley drawings.

After the communist period, all camouflaged uniforms were replaced with two types of patterns. A woodland type with shapes more or less similar to the U.S. leaf pattern but markings present green and white clouds. This uniform has been worn for example by the 36th Navy Infantry (Marines), named Vasile Lupu. A recent pattern worn during European maneuvers was made of small brown spots with larger black and green shapes on a yellow gray background. This new camouflage was recently worn in Central European maneuvers. It was later also worn by infantry regiments and border guards. In 1999 there appeared an urban camouflage for police, similar to woodland but made of black, gray, blue and white shapes.

Close up view of the new Romanian Woodland camouflage. Author's collection.

Romanian soldiers in 1998 in maneuvers, red beret. NATO photo service, Brussels, Belgium.

Exercise Cooperative with Romanian troops Osprey 1996 with NATO armies. Camp Lejeune, Jacksonville, USA, August 1996. Second pattern. NATO photo service, Brussels, Belgium.

LEFT: NCOs and officers of the Romanian army in maneuvers. Second pattern 1998. Y.D. via archives.

A Romanian police soldier in blue uniform with a girl friend in red. Reserved rights.

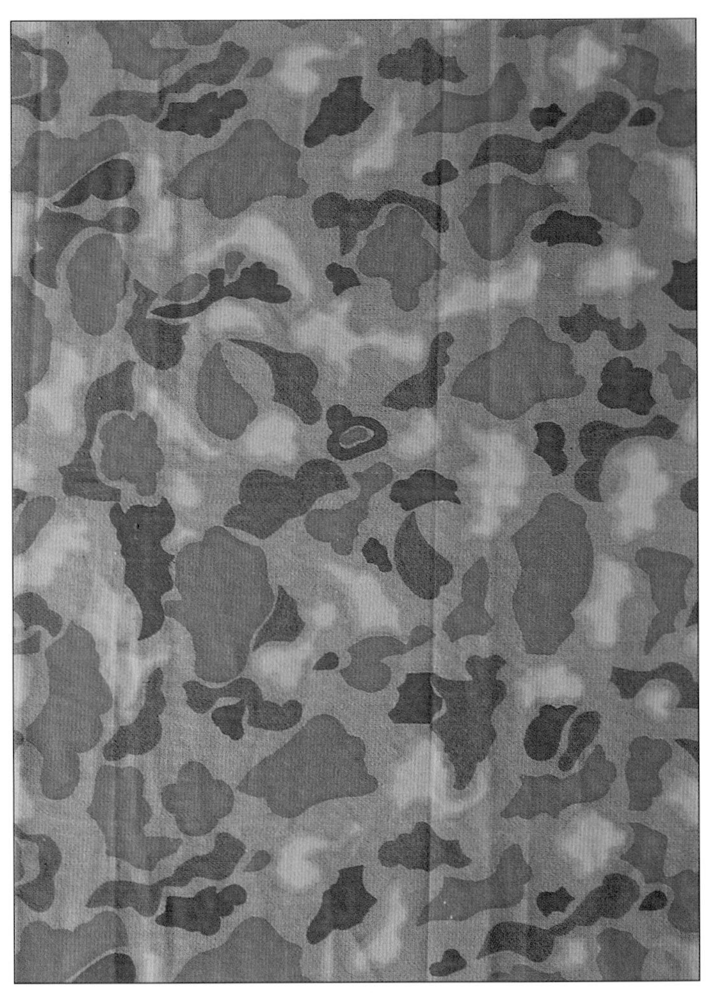

Close up view of the Turkish Turquoise blue pattern. For cover helmets and quarter shelters, the drawings were green and not blue. Author's collection.

"Turquoise blue" of the 1970 Turkish army. Similar to U.S. M44 King Kard but with variants of blue drawings. Wild mook 49, Japan.

1980 new Turkish camouflage similar to U.S. M65-67 ERDL made in Turkey. Y.D. via archives.

Close up view of the Turkish pattern similar to U.S. M65-67 ERDL.

Brown camouflage with U.S. M65 drawings but made in Turkey. This pattern is also very similar to the brown Slovakian pattern and Nicaragua 1970. Y.D. via archives.

Close up view of brown Woodland. These pale colors are similar to the Nicaraguan and Slovakian patterns.

New 1996 Turkish pattern. The drawings are longer and horizontal, very different from the 1980 pattern. Y.D. via archives.

Turkey

The famous "turquoise blue" is no longer much worn and used only as a reversible ground sheet or waterproof poncho tent. In 1980 the Turkish army adopted a similar pattern to the U.S. leaf, but made in Turkey with another variant, brown with white twigs. Some helmet covers were made of frog spots, green instead of turquoise blue. Turquoise blue pattern has been used in 1974 "Attila" operation in Cyprus. The new uniforms have been worn in Yugoslavia in 1994 and also for elite troops "Parasutai Piyade"

Ukraine

The Ukrainian army could be seen during maneuvers with Polish and American troops in 1996 wearing the U.S. M80 woodland camouflage uniform.

In Bosnia, the battalion 240 of the 17th armored division 1996 and 1st paratrooper division wore the Russian camouflage pattern.[3]

This Ukrainian cap with Swedish splinters pattern is adorned with a cockade with the symbol of Ukraine, the three-toothed fork (trident of St Vladimir). In 1996 during maneuvers of NATO Ukrainian soldiers wore U.S. M81 Woodland. The Ukrainian ministry never replied to our requests.

Notes:

[1] See Desmond, Dennis. *Camouflage Uniforms of the Soviet Union and Russia, 1937-To The Present,* pp.145, 146.
[2] id.pp.118, 123.
[3] id.p.122.

Ukrainian soldier in UNO former Yugoslavian troops in 1995. Russian camouflage. Y.D. via archives.

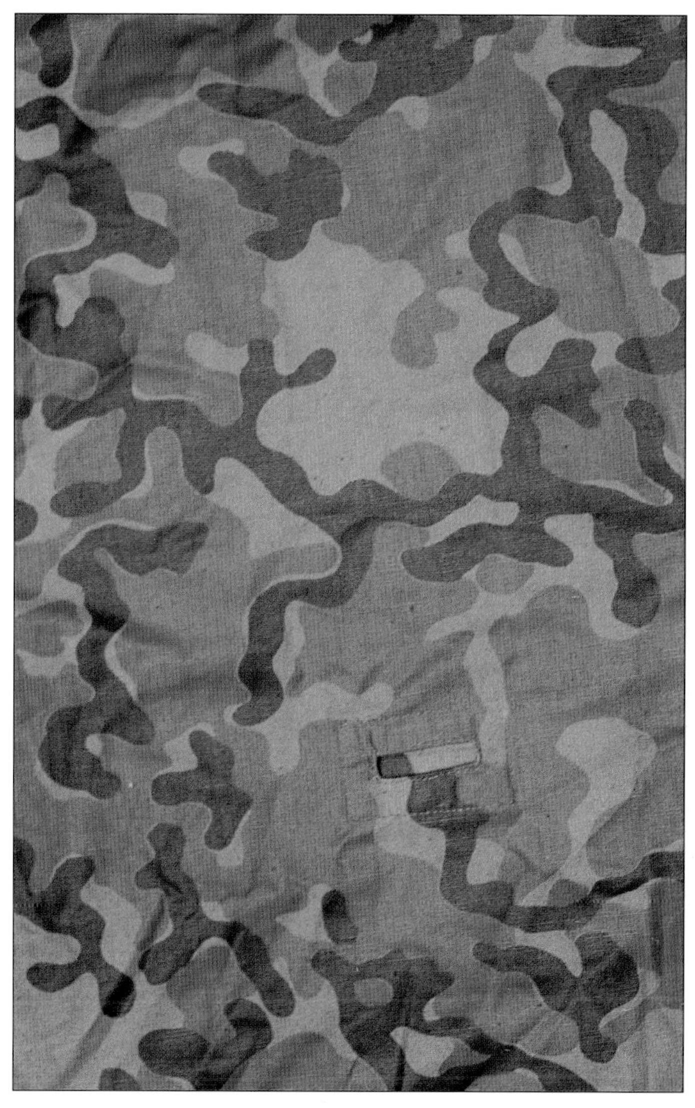

Close up view of Russian camouflage, dark green "worms" for Ukrainian troops at the beginning of the country's independence.

6

Camouflage Uniforms of Canada, USA and Miscellaneous

Canada
During many years the Canadian army wore British camouflage DPM. In 1997 a project for a new suit for combat was tested, and in 1999 this camouflage has been distributed to limited units. It is designed like the U.S. army M81, but the drawings are shorter, and the shades are very clear and composed of green dark and black spots.

Canadian forces, seen here in 1970, have worn the British DPM since 1960. Private collection.

This Canadian project has been at trial for two years and will probably be adopted in 1999. Courtesy of Canadian photo service in Ottawa.

The 1999 camouflage pattern of Canada, which is made of very clear green, black, or dark green shapes. Courtesy of Canadian Ministry of defence in Ottawa.

United States of America

Experiments in camouflage uniforms by the U.S. army started in 1940 at the Army Corps of Engineers. The famous "frog" pattern was created by a horticulturist and gardening editor for "Better Homes and Gardens." This pattern, made of rounded shapes, two faces (one green and the other brown) was issued in 1942 and mainly supplied for the Pacific campaign against the Japanese. It was briefly provided to soldiers during the Normandy landing in 1944.

For the Asian jungle this material proved highly unpopular, since it was too hot and awkward for the jungle. The marines, instead of it, adopted a two piece utility suit that was better than the one piece overall formerly distributed, and this pattern became famous and was imitated by many foreign armies after the war.

In 1960 there was no official camouflage uniform in the U.S. army. During the Korean war a test had been made with the "leaves and twigs" pattern, but was not widely worn and was soon dropped.

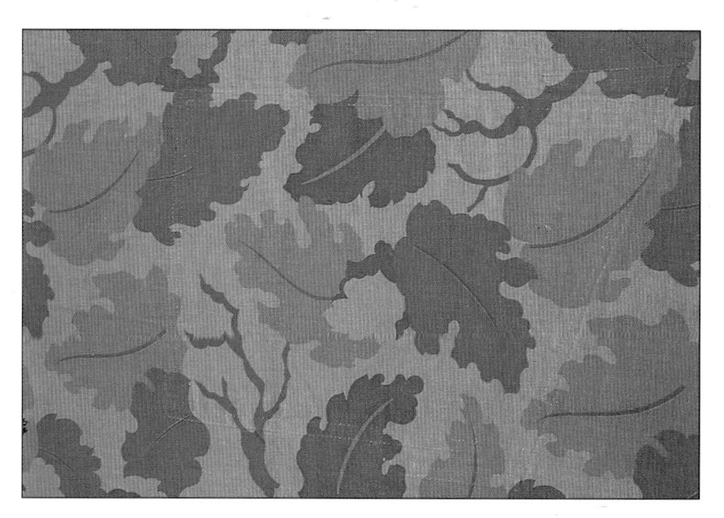

Close up view of "wine tree leaves" pattern with ochre red twigs. The helmet cover was reversible with variant browns, and marked: DSA 100 74F. U. 919. MPLS C° Soc. F/T Blind inc. Author's collection.

Only helmet covers of this pattern were used. Note that this pattern was reversible, the other side being made of different brown shapes called "brown clouds."

U.S. military advisors arrived in SVN in the early 1960s, and special forces serving with irregular Vietnamese troops were provided by the CIA with "duck hunter" camouflage uniforms, based on a WWII U.S. pattern and sold by commercial suppliers, such as Sears Roebuck.

As the conflict in Vietnam grew, the need for camouflage was expanded. U.S. military advisors turned to the Vietnamese army for inspiration. The Vietnamese Marine Corps indeed had a special camouflage suit since 1959 consisting of horizontal shapes printed over a dark and light green background, the famous Tiger stripes pattern. This pattern was derived from the French army brush strokes, named the French Lizard pattern of 1953. The Tiger stripe is at the present time used by certain units of the Russian Army and

The U.S. trial for the Korean war in 1950. Very rarely worn. The cover helmet would be maintained in South Vietnam until 1965 and later. Author's collection.

New camouflage "leaf pattern" for U.S. troops in South Vietnam until 1975. U.S. M65 ERDL BDU. Author's collection. Transparency : Photo Violet, Paris.

Three patterns used by U.S. troops in South Vietnam. At left M65 BDU, at right two sorts of "tiger stripes" worn by U.S. special forces and military advisors in the SVN army. Author's collection.

The U.S. M 81 Woodland: the drawings are larger than U.S. M65. Markings: DL 100. 81. C. 2413. Selma apparel Corp. Marines in Camp Lejeune, USA, 1996. NATO photo service, Brussels, Belgium.

U.S. M 81 Woodland material, close up view. Author's collection.

Some trials of the U.S. Army since 1975: rain strokes, squares, desert pattern. Dan Peterson collection.

Desert pattern "chocolate chips" markings: Mil. T 43217. DLA 100. 82. S 3100. SW1 1982 USA.

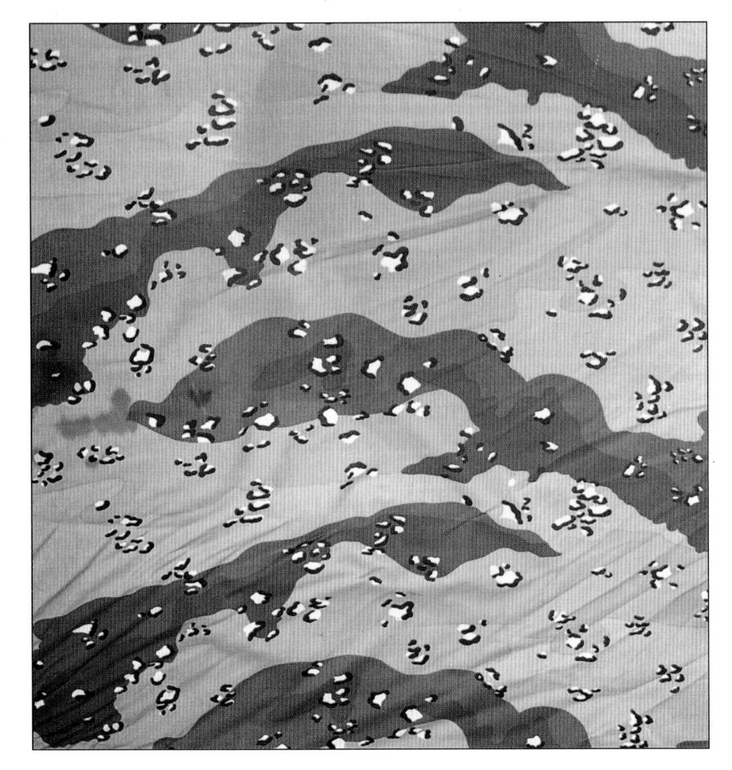

Close up view of horizontal drawings of the desert daytime U.S. pattern.

A group of U.S. soldiers in the "chocolate chips" pattern. Y.D. via archives.

Desert night camouflage parka trialed in 1989-90 during the Gulf war. Markings: DLA. 8C. 2532 Lancer clothing corp.

Close up view of the rip stop desert pattern of U.S. Army in 1998, new pattern.

has been much studied by the American collector Richard Johnson (SLC Utah).1

A special American pattern, the M65 Leaf pattern issued in 1965, was reserved in 1967 for elite units of the U.S. army in Vietnam. In 1970, inspired by the Vietnam experience, American research into camouflage continued and created in 1981 the Woodland pattern. The Woodland is very similar to the M65 and M69 ERDL leaf pattern, but drawings and shapes are much larger.

During Operation Bright Star the 82nd U.S. army and Egyptian forces organized a combined exercise in North Africa, and two desert patterns were tested. A chocolate chips with long brown stripes and white pebbles edged with black shading lines became accepted as the daytime desert pattern. The night desert pattern had a dark green and clear green checked pattern with clusters of little black squares. This uniform is twenty percent less visible at ranges less than 100 meters, and invisible at 200 meters or more. All contain anti-infra red dyes.

Since 1992 another desert pattern has been created with little brown stripes. All studies and tests have been realized by the engineering technical center of Natick in Massachussetts under the direction of Mrs Commerford and her group of specialists and technicians.

The new desert pattern of the U.S. Army in 1991 and adopted for all troops in 1996. Worn only by General Schwarzkopf during Gulf war '89-90.

Factories Labels

All camouflage uniforms we have been able to collect for more than twenty years have not always had a label inside. Often we found them in far away flea markets, old stores, or even through friends living in foreign embassies or working in military organizations overseas. So we collected more than three hundred different patterns, real uniforms or only good pictures of them. We will present all the labels we have been able to select, since they are useful in case of doubt as to the real origin of the uniforms.

These markings have been photographed inside camouflage uniforms, but they are different according to factories and time of manufacturing. For example, for French camouflage uniforms, some of them are made in Toulouse, while others are made in Saintes, so the labels are different.

Some manufacturers, having sold their ordered uniforms to the army, took advantage of fashion and craze for this type of pattern. So it is possible to buy civilian uniforms made for hunting or camping and believe they are military uniforms. Our collection is made only of genuine military material, and most of the official labels have been presented in this book. (*See following photographs for corresponding labels)*

1 German "desert areas"
2 Romanian "parsley" pattern
3 Czech "rain strokes"
4 German "woodland"
5 Democratic German "leaves"
6 Polish large colored shapes
7 Finnish all patterns
8 Swedish "splinters"
9 Belgian "leibermuster"
10 Spanish "brown puzzle"
11 French "quarter shelter"
12 Italian "San Marco"
13 Old Dutch pattern
14 Hungarian "ground sheet"
15 Cuban pattern for Serbs 1
16 Cuban pattern for Serbs 2
17 Russian "KGB leaves"
18 French "lizard"
19 Russian "MVD leaves"
20 Czech caps
21 Desert British DPM
22 Swiss "leibermuster"
23 Polish "little strokes"
24 Hungarian "brown bowls"

25 Desert British (Gulf War)
26 Polish 2nd type of "strokes"
27 French "quarter shelters"
28 U.S. army cover helmet
29 King Kard US HBT 1943
30 Urban Camouflage U.S.
31 British camouflage 1943
32 Desert camouflage U.S.
33 Modern Spanish patterns
34 Italian camouflage "quarter shelters"
35 Bundesgrenschutzen uniformen RFA 1960 (German customs)
36 MacLaren Sportswear Company trial for a 1943 U.S. camo uniform
37 The M81 Woodland U.S. camouflage
38 Austrian 1958-78 camouflage pattern
39 Half tent of Korean U.S. camouflage 1963
40 East Germany camouflage 1963
41 British camouflage smock

Influences of WWII and U.S. Army

Many countries have been greatly influenced post-WWII by '37-45 German experience and studies in the field of camouflage uniforms. Eastern Warsaw Pact armies, such as Bulgaria, Poland, and Czechoslovakia have created similar patterns to Third Reich armies, and even a small number of Western European, such as Belgium, Federal Germany, and Switzerland have also mimicked these camouflages.

But the studies of the U.S. army, mainly carried out in Natick engineering center, Massachusetts, have modified this orientation and at present time many European armies have made patterns similar to the U.S. or ordered directly from U.S. army supplies, Woodland M81 camouflage uniforms (mainly new and smaller armies, such as Albania, Latvia, Lithuania, Croatia, etc.)

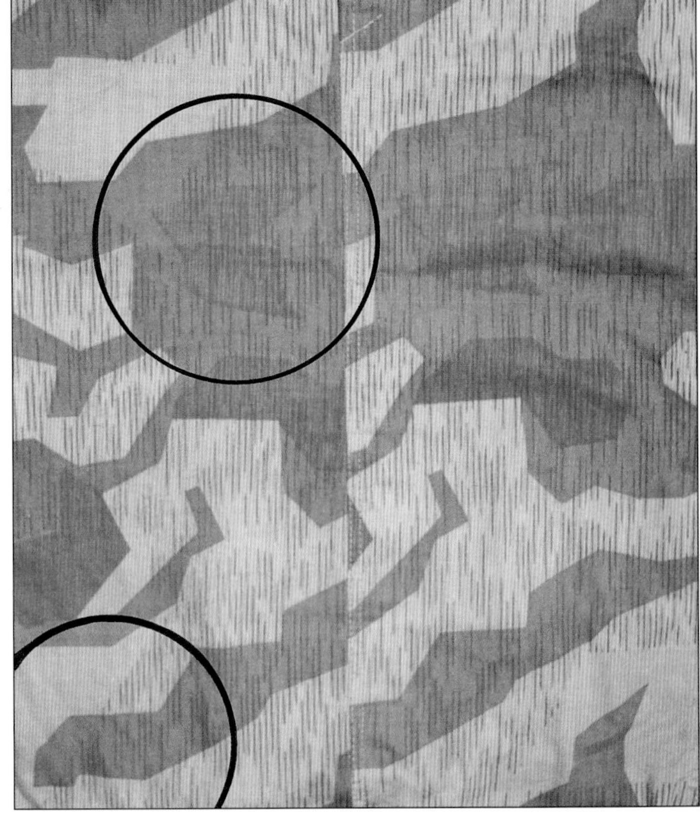

Splinter camouflage of WH 37-45. The drawings are the same as Swiss pattern 1947-55.

German Gebirgsjäger with WH splinters 1937-45 on winter padded parka. ECPA archives, France.

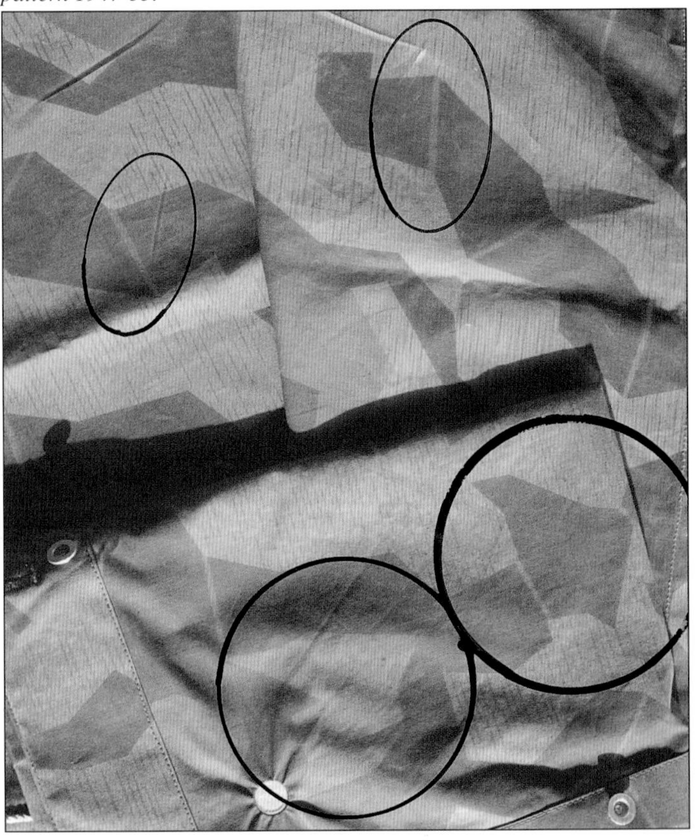

The drawings of Swiss patterns are the same as German 1937-1945. J.P. Soulier collection.

German soldier of artillery with the zeltbahn 1937. A circle shows a very special drawing, exactly the same as the Polish camouflage pattern 1970.

German paratrooper with jump suit wearing the 'splinter special pattern for Luftwaffe.

A Polish soldier with hood parka in 1970. Splinters have different colors, but the drawings are the same.

Typical splinters of Luftwaffe.

ABOVE LEFT: A French ground sheet pattern 1940 has been printed with German paratroopers 1941 pattern maybe for trial and to observe efficacy of the drawings, post war.

ABOVE RIGHT: Leibermuster 1945 last German trial of third Reich. It was undetectable in infra red camera research. Markings: RZ N° 60 / 0135 / 5043. Author's collection.

A Bulgarian soldier wearing similar splinters to the German paratroopers' pattern with rain strokes.

Swiss pattern, influenced heavily in 1960 by German leibermuster.

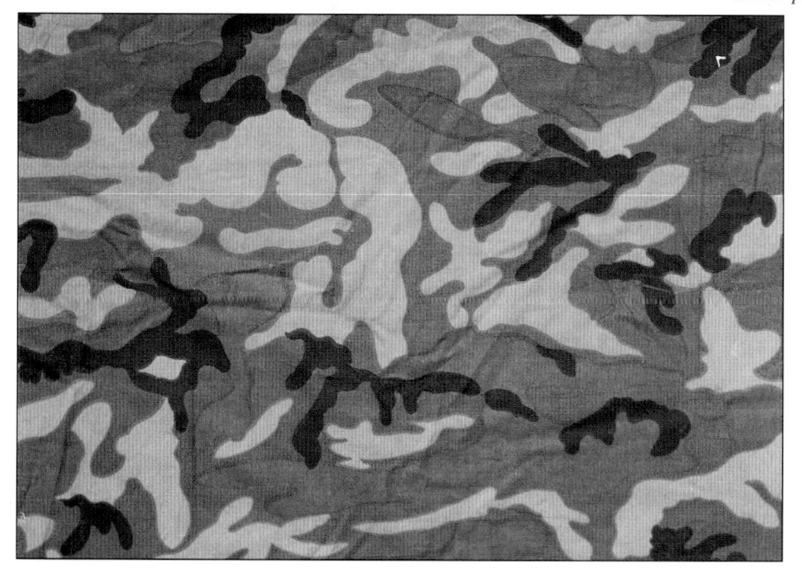

ABOVE LEFT: Belgian camouflage trial 1955 ABL KH, similar to leibermuster.

LEFT: A copy of U.S. Army M67 or M81 (Spanish pattern after 1982).

A copy of U.S. Army M67 or M81 (Italian pattern).

A copy of U.S. Army M44 King Kard 43-44, generally called "Duck hunter," and imitated by many armies in the world until 1970.

Life and Death of Camouflage Uniforms

The period of printing and drawings to conceal a man in the field of battle is over. Since 1944, detection of soldiers walking in the jungle by night has been an important problem in the Pacific. So, U.S. engineers, a few years after Germany, have studied the best means to detect the enemy.

In February 1945, the first camouflage suits not detectable with infra red cameras were distributed to a group of soldiers in the German army. The light material, made of 100% cotton, was impregnated with carbon black at 4%, so in a forest, on a chlorophyll background, the carbon black becomes red, exactly like chlorophyll. In July 1945, Lieutenant Richardson of U.S. army headquarters introduced in different studies of German material found in stocks a report on infra red possibilities to detect targets by night, and also all camouflage uniforms.[2]

The studies had indeed started in Europe and were rapidly increased by many countries, mainly Sweden (Barracuda establishments) and the U.S. In France, Texunion factories published in 1976,

The "Ghilly II," or different sorts of plastic leaves sewn together. They can decrease smell of sweat and escape infra red detections.

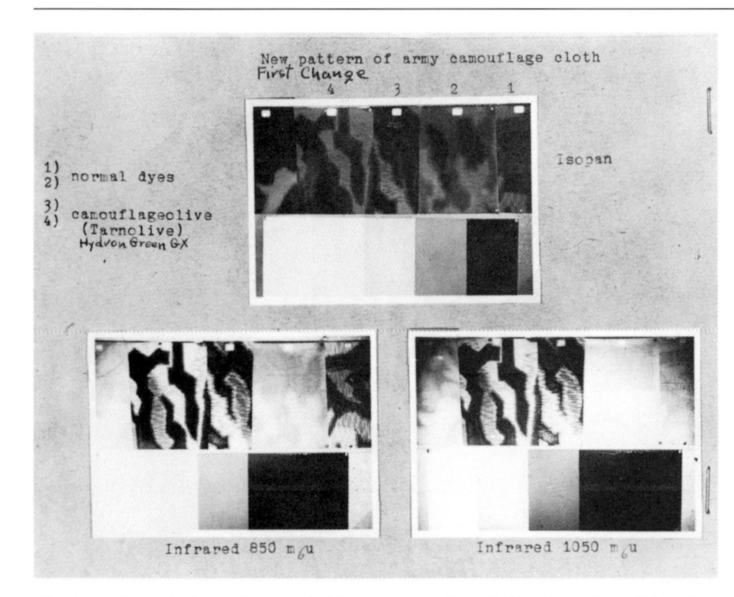

Engineering trials with material impregnated with black carbon. The photos, taken with an infra red camera, show that the material is undetectable. Richardson report QMC 1945.

In nature, sites are ready to receive men or vehicles (sites are visible on the first photo, in circles).

with German studies of Marquardt and Schulz, a report on infra red detection, and a special cloth undetectable with infra red cameras.

New Conception of Camouflage Patterns

The effect of camouflage is achieved through changing the form, color and contrast of essential objects in relation to their surroundings. Therefore, the camouflage must be effective in all those parts of the electro-magnetic spectrum within which one can expect to be observed.[3] It is also essential that the camouflage is efficient at all viewing ranges. If we study a forest at a close range it appears to consist of a very large number of small units of different shapes and colors and of varying brightness. Farther off these seem to melt

In the second photo, sites are camouflaged with tarpaulins impregnated or not with black carbon. The camouflage is effective.

In the third photo, taken with an infra red camera, the sites stay camouflaged if the material is impregnated with black carbon. Barracuda camouflage establishments photos, 1975.

Photo showing a German soldier in woodland with field jacket and cover helmet impregnated with black carbon. The 1942 material is genuine. Courtesy of J.P. Soulier.

into larger units, and we perceive only large areas of varying brightness. The resolution is no longer sufficient for distinguishing the small units.

The faculty of color vision also changes. At long range, colors appear to be less clear, different color shades melt together, and the intensities of colors weaken. A good camouflage must, of course, produce the same effect. The pattern and color effect must be designed with consideration to both the close range and the distant effects they produce. Modern camouflage material is far superior to older types in these aspects.

Good camouflage begins with a well thought out design and coloring of the war material used. This is particularly important, as external camouflage in many cases cannot be used during combat or transport. The coloring must be well adapted to the expected battle environment, e.g. through pattern painting according to modern methods. Hot surfaces should be shielded so that they cannot be detected by thermal reconnaissance.

Different objects reflect amounts of light in varying wavelength ranges. This gives them their brightness and color. The following formula applies to opaque objects: the emissivity + the reflectivity = 1. Thus, an object with high emissivity has low reflectivity and is consequently dark. The table below gives the approximate reflectivity for various earth surfaces in the visible part of the spectrum.

Surface: Snow, reflectance: 0.70-0.86
Surface: Clouds, reflectance: 0.50-0.75
Surface: Limestone, reflectance: 0.63

Surface: Dry sand, reflectance: 0.24
Surface: Wet sand, reflectance: 0.18
Surface: Bare ground, reflectance: 0.03-0.20
Surface: Water, reflectance: 0.03-0.10
Surface: Forest, reflectance: 0.03-0.15
Surface: Grass, reflectance: 0.12-0.30
Surface: Rock, reflectance: 0.12-0.30
Surface: Concrete, reflectance: 0.15-0.35
Surface: Blacktop roads, reflectance: 0.08-0.09

The human eye plays a most important role in the majority of reconnaissance systems. It can either be used for direct observation, or be magnified with the help of binoculars or an image intensifier. Moreover, it can be used for viewing photographs or displays. It is therefore important to understand the eye's function and to know its limitations. The sensitivity of the eye varies continuously with the wavelength from a maximum somewhere in the middle of the visible spectrum down to zero at its two ends.

As there are two types of retinal receptors in the eye, rods and cones, the eye's spectral responsivity in the dark adapted (scotopic) state is different from that in its light adapted (photopic) state. In daylight conditions, when the luminance of the visual fields is at least 1 cd/m^2 (cd = unit of luminance = Candela/m^2), the cones are fully functional and we have the ability to see colors. At low levels of illumination, i. e. below 10^{-4} cd/m^2, when only the rods function we have no such ability. In between we have the mesopic range where the vision continuously changes from the photopic to the scotopic state. With decreasing luminance of the visual field, the eye's ability to perceive small objects that have little contrast against their background also decreases.

Study of Mr. Piet Bess on infra red detection. Photos Tony Moffitt, Military Illustrated p.10 to 14 and 15 to 18, courtesy of Tim Newark.

Mr. Bess shows two pictures in which French Army and U.S. Army uniforms are photographed with an I.R. camera. No comments.

Direct observation has many important applications in the field of reconnaissance. It can be by means of binoculars during reconnaissance on the ground, or unaided for ground reconnaissance during air raids. Direct observation has many advantages. It offers immediate information on which action may be based; the picture is seen in true third dimension and is easily evaluated by the brain. The eye is normally an accurate and sensitive receiver and allows observation of movement. It has at least four major disadvantages: there is no permanent record for future direct comparison, weather and time of day may limit its performance, the observer's experience and mission may limit the information obtained, and human error may result in incomplete and incorrect information.

For the experts, chlorophyll is of great interest to camouflage due to the fact that it impacts green coloring to vegetation within the wave length range perceptible to the human eye. Other plant substances are responsible for reflections at wavelengths just beyond the visible spectrum. See the numbers of reflectance and surfaces.

Since 1975, camouflage has become a real science, and the old printed material with different colors made for the deception of enemies' eyes is obsolete in 1999. Detection has become a grand art, and includes new words and phrases: furtive material, reduction of radar signature, thermic snares, "windows," jam radios, and so on.

Soon, old camouflage uniforms will die, maintained only in smaller nations without possibilities to use such sophisticated materials. A last there is hope for lovers of this kind of garment: in Lympstone, U.K., a special uniform is being tested by elite commandos of Royal "Marines," known as the chameleon suit. This material is made to replace the already obsolete Ghillie II. It absorbs sweat smell, is undetectable by infra red cameras, and includes plastic leaves which reproduce the exact sound of "wind in a woodland" (devised by Karrimor society).

Maybe one day, somebody will create a material with the chameleon system, which also reproduces colors of a direct background, as so many fishes and mollusks know how to do it.

Notes:

[1] See Johnson, Richard. *Tiger Patterns: A Guide to the Vietnam War's Tigerstripe Combat Fatigue Patterns and Uniforms*. Atglen, Pennsylvania: Schiffer Publishing Ltd., 1998.

[2] Camouflage fabrics both plain and printed for military use by the German SS and German army, reported by Francis S. Richardson, QMC consultant 20th July 1945.

[3] According to a study of Barracouda Camouflage Department, Djurshölm and Lahölm, Sweden.

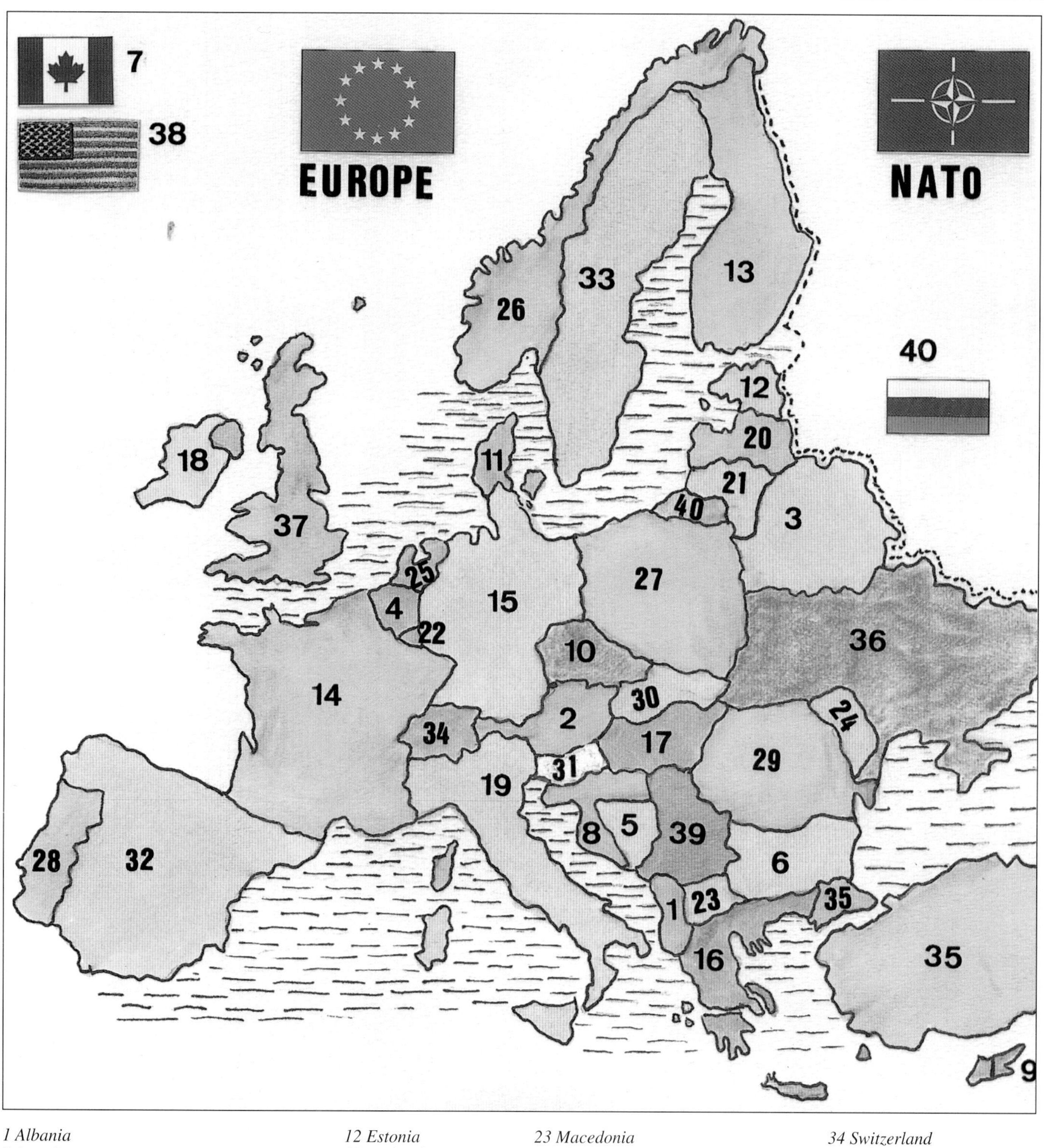

1 Albania
2 Austria
3 Belarus
4 Belgium *
5 Bosnia
6 Bulgaria
7 Canada*
8 Croatia
9 Cyprus
10 Czech
11 Denmark*
12 Estonia
13 Finland
14 France*
15 Germany*
16 Greece*
17 Hungary
18 Ireland*
19 Italy*
20 Latvia
21 Lithuania
22 Luxembourg*
23 Macedonia
24 Moldavia
25 Netherlands*
26 Norway*
27 Poland
28 Portugal*
29 Romania
30 Slovakia
31 Slovenia
32 Spain*
33 Sweden
34 Switzerland
35 Turkey*
36 Ukraine
37 United Kingdom*
38 United States of America*
39 Yugoslavia
40 (Russia)

*NATO forces

1
2
3
4
5
6
7
8
9
10
11
12
13
14
15
16
17
18
19
20
21
22
23
24
25
26
27
28
29
30
31
32
33
34
35
36
37
38
39
40

Bibliography

Selected Books and Articles

Albania:

Shtepia Botuese. *Ushtria Popullore*. Tirana: 8 Nentori, 1986.

Austria:

Urrisk, Rolf. *Die Uniformen des Osterreichischen Bundesheeres.* Graz: H. Weisshaupt Verlag, 1994.

Belgium:

Smeet, J. *Amilitaria Magazine*. Brussels: New Fashion Media, November 1983, pp, 70-72.

Bulgaria:

Na Straje Mira i Sotzialisma. Moscow: Planeta, 1985.

Cyprus:

Debay, Yves. *Commandos Chypriotes*, *Raids*, N° 12O, Paris: Ed H et C., 1996.

Croatia:

Tudman, Ankika. *Hrvasto Ratno Znakovlje, Kniga 1*. Zagreb, 1995.

Czech:

Gothard, Vl. and Loucka, M. *Nove Stejnokroje v Armada Ceske Republiky.* Praha: Ministerstvo Obrany, 1996.

Finland:

Information division Defense Staff. *Facts about the Finnish Defense Forces*. Helsinki: Edyta Oy Ltd., 1998.

France:

French Camouflage Uniforms, Terre Magazine. Paris: SIRPA, Ministère de la Défense, 1994 to 1996.

Greece:

Hellinikos Stratos, Quatermaster monography. Athens: Defense Ministry, 1994.

Germany:

Fleisher, W. and Eierman, R. Tarnen, *Taüschen und Attrappen.* Wolfersheim: Podzun Pallas, 1998.

Na Straje Mira i Sotzialisma, Moscow: Planeta, 1985.

Hungary:

Drawings and plates of Hadtortenety museum. Budapest, 1996.

Ireland:

Staff and Cosentoir, *Defense handbook*. Dublin: Austin Pender Ed., 1988.

Italy:

Viotti, A. *Uniformi Distintivi dell' Esercito Italiano*. Rome: Stato maggiore dell esercito, Ed. Ufficio storico , 1988.

Latvia:

Latvijas Kara Musejs booklet. Riga, 1998.

Luxembourg:

D'Armei Am Dengscht Vun Den Menschen. Luxembourg: Ministère de la Force publique, 1998.

Poland:

Sawicki, Z. and Waskiewicz, J. and Wielechowski, A. *Mundur Wojska Polskievo*. Warsaw: Wydawnictvo Bellona, 1997.

Romania:

Sladowski, M., Babiuc, V., Degeratu, C., Costache, M., Zaharia, D., Diaconescu, G., Ghitas, G., Botezatu, P., Teodorescu, C., Dudu Ionescu, C. *NATO Sixteen Nations*, vol. 42, special issue. Germany and Netherlands: J. Perels Publishing C° Mönch publ. group, 1997.

Slovakia:

Hollij (pplk) and Pekarik (pplk). *Uniformy Armada Slovenskej Republiky*. Bratislava: Ed. Bucher Printex , 1998.

Slovenia:

Svajncer, J. (major) *Territorialna Obramba Republiky Slovenije*. Lubliana: Ed. Univerzitetna Knijznica ,1992.

Spain:

Bueno, J., Gravalos, L., Calvo, J.L., Ejercito E*spanol Uniformes Contemporaneos*. Madrid: Ed. San Martin , 1980.

Sweden:

Barracuda factory Lahölm, booklet. Stockholm: Högkvartett info S 107, 1970.

Switzerland:

Swiss Army Regulation. Bern: Abteilung Ausrüstung, 1986.

United Kingdom:

Tanner, J.K., Cox, Michael. *British Forces in the Gulf*. London: Military Illustrated, N°27, pp, 24-29.

Smith, D., Chappell M. *Army Uniforms since 1945*. Poole: Blandford Press Ed., 1980.

Leroy Thompson, Chappell, M. *Uniforms of Elite Troops*. Poole: Blandford Press Ed., 1982.

United States of America:

de Barba, P. *Les Forces Speciales US dans la Guerre du Golfe. Raids Magazine*, N°96, Paris: Ed. H et C, 1998.

Desmond, Dennis. *Camouflage Uniforms of the Soviet Union and Russia, 1937-To The Present*. Atglen, Pennsylvania: Schiffer Publishing Ltd., 1998.

Itaya, Jun. *US Army Combat (1)*. Tokyo: Kesaharu Imai Ed., 1982.

Kikuchi, M. *US Marines Corps To Day Combat (6)*. Tokyo: Kesaharu Imai Ed., 1982.

Leroy Thompson, Chappel, M. *Uniforms of Elite Troops* . Poole : Blandford Press Ed., 1982.

Leroy Thompson. *Uniforms of the Indo-China and Vietnam Wars*. Poole: Blandford Press Ed., 1984.

Moran , J. *US Marines Uniforms and Equipment*. London: Windrow and Greene Ed., 1992.

Merenyi, Kathryn. Harris, Gary. *Uniforms of the Armed Forces of Eastern European Armies*. Defense Intelligence Agency.

Yugoslavia:

Debay, Y. *Printemps de Guerre en Bosnie. Raids Magazine*, N°107, Paris: Ed. H et C, 1995.

Furlan, M. Bjelos, H. *Army and Insignias*. Toronto: Militaria House Ed., 1985.

Ministry of Defense. *Vojcka Moderna Iabna Retch*. Paris: Ed. Université de Paris, bibliothèque contemporaine, 1995.

Magazines and Militaria Catalogs

American Survival Guide, Mac Millen publ. Po Box 15690, Santa Ana ,CA, USA.

Archeologia Wojskowa Militaria, Burdzskego st. 9/5, Warsaw, Poland.

Armada International, Thurgauerstrasse 39, Zurich, Switzerland.

Armées d'Aujourd'hui, ADDIM 6 Rue St. Charles, 75015 Paris, France.

Armées et Defense, 172 Av. Marie José, 1200 Brussels, Belgium.

Armies Badges of Elite Troops, Bruce Quarrie, London, United Kingdom.

British Soldier in the 20th Century, Wessex Military Publishing, Po Box 19, Okehampton, Devon, United Kingdom.

Catalogue US cavalry, Centennial Av. Radcliff, Kentucky, USA.

Catalogue U.S.M.C., 90 rue de La Folie Méricourt, 75011 Paris, France.

Combat Suits Catalogs Wild Mook, 1 to 50, World Photo Press, Kabuki Cho, Tokyo160, Japan.

Defensa, Madrid, Spain.

Defense et Armements, 15-17 Quai de l'Oise, 75019 Paris, France.

Deutsches Waffen Journal, Germany.

Ejercito Espanol, Ediciones San Martin, Madrid, Spain.

Elite Troops Series, Osprey London, United Kingdom.

Front Yugoslavian Army Magazine, Po Box 36, Belgrade, Yugoslavia.

Gazette des Uniformes, Regi Arms, 15 rue Cronstadt 75015 Paris.

Hachette Series Uniformes, Connaissance de l'Histoire, Paris, France.

Heraldry and Regalia of War, Phoebus Press, London, United Kingdom.

Images de Guerre, ALP Cavensish , rue de La Rochefoucauld, 75009 Paris, France.

Irish Defense Forces, Army Headquarters, Dublin, Ireland.

Jane's Defence Weekly, Sentinel House, Horley, Surrey, United Kingdom.

Military Illustrated, 43 Museum Street, London, United Kingdom.

Militaria Magazine, 19 av. de la République, 75011 Paris, France.

PX Militaria, World Photo Press, Kabuki Cho, Tokyo160, Japan.

Raids Magazine, Histoire et Collections, 5 av. de la Republique. 75011 Paris, France.

Rivista Militare Italiana, Plazza G. Alessi, Genova, Italy.

Rivista Difesa, Chiavari, Italy.

Storia Militare, Uniformi Tuttostoria, Via Sonnino, Parma, Italy.

Trente Neuf Quarante Cinq Magazine, Heimdal, 14400 Bayeux, France.

Uniformi (See to *Storia Militare*).

Wessex Military Publishing, Po Box 19, Okehampton, Devon, United Kingdom.

Author's Books And Articles
Modern Period

Camouflage Uniforms of Warsaw's Pact and Non Aligned Armies in Europe. London: Iso Galago Publishing, 1989.

Camouflage Uniforms of Nato, London: Iso Galago Publishing, 1990.

Camouflage Uniforms of Gulf War and Northern African Armies. London: Iso Galago, Publishing,1989.

Camouflage Book. With T and Q. Newark London: Brassey's Ltd., 1996.

Camouflage Uniforms articles in Militaria Magazine, France. Gazette des Armes, France. Amilitaria, Belgium. Jane's Defence Weekly, United Kingdom. Survival Guide, USA. Rivista Italiana Defensa, Italy.

1939-1945 Period

Camouflage Uniforms of WSS. Part 1. London: Iso Galago Publishing, 1986.

Camouflage Uniforms of WSS. Part 2. London :Iso Galago Publishing, 1988.

Camouflage Uniforms of Wehrmacht. London: Iso Galago Publishing, 1988

US Richardson Report 20 July 1945. London: Iso Galago Publishing, 1989.

Tenues Camouflées de la 2ème Guerre Mondiale. Paris: Regi Arms, 1992.

Les Parachutistes Allemands. Paris: Regi Arms, 1993.

Camouflage Uniforms of Waffen SS, Beaver, M. with Borsarello, J.F. Atglen: Shiffer Publishing, 1995.

also from the publisher

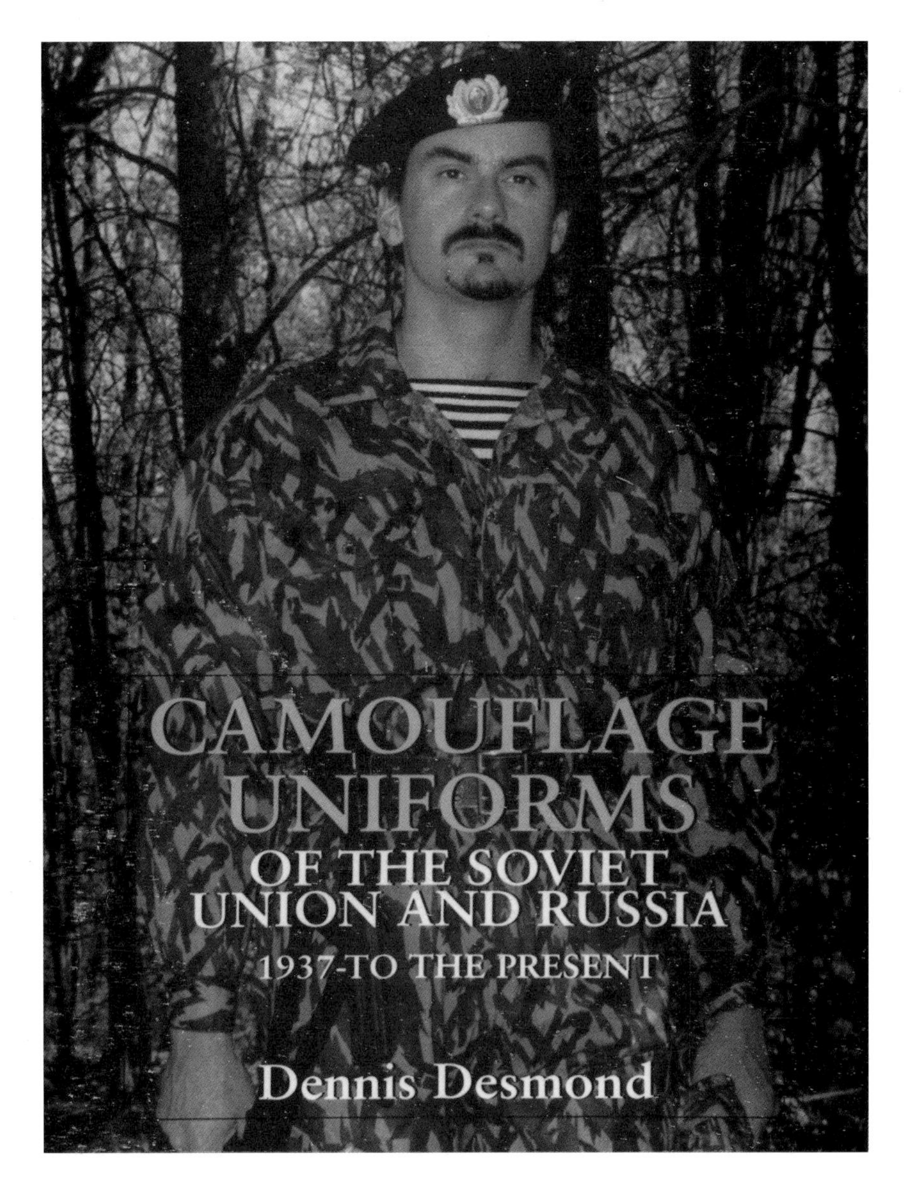
CAMOUFLAGE
UNIFORMS
OF THE SOVIET
UNION AND RUSSIA
1937-TO THE PRESENT
Dennis Desmond